GW00792359

STARS OF THE SOUTHERN SKIES

Patrick Moore has been interested in astronomy since the age of six. During World War II he flew as a navigator with RAF Bomber Command, and subsequently established his private observatory in Sussex. He was a pre-*Apollo* mapper of the Moon, and is the author of many technical papers as well as popular books. His television series on the BBC, *The Sky at Night,* has been broadcast every four weeks since April 1957 — easily a world record; Patrick Moore has presented all of them.

He is a member of the International Astronomical Union; a Fellow of the Royal Astronomical Society (Jackson-Gwilt Medallist, 1985); a Council Member of the British Astronomical Association (President 1982–4; Goodacre Gold Medallist 1988); an Honorary Member of the Royal Astronomical Society of New Zealand and of the Royal Astronomical Society of Canada; and also an Honorary Member of the Astronomical Society of the USSR. He was the Roberts-Klumpke Medallist of the Astronomical Society of the Pacific in 1987. He was awarded the OBE in 1966 and the CBE in 1988 for his services to astronomy.

STARS OF THE SOUTHERN SKIES

PATRICK MOORE

Published in 1994 by David Bateman Ltd,
"Golden Heights", 32–34 View Road,
Glenfield, Auckland, New Zealand

Distributed in southern Africa by
Struik Distributors (Pty) Ltd
P.O. Box 624
Bergvlei 2012
South Africa

Copyright © Patrick Moore, 1994
Copyright © David Bateman Ltd, 1994

ISBN 1 86953 185 X

This book is copyright. Except for the purpose of fair review, no part may be stored or transmitted in any form or by any means, electronic or mechanical, including recording or storage in any information retrieval system without permission in writing from the publisher. No reproduction may be made, whether by photocopying or by any other means, unless a licence has been obtained from the publisher or its agent.

Design by Errol McLeary
Cover design by Cathy Larsen
Cover photograph by Ronald Royer
Typeset in 10/11 Trump Medieval
Printed in Hong Kong by Colorcraft

Front Cover: The southern hemisphere night sky showing the glorious star-clouds in the constellation of Sagittarius.

Contents

Foreword

In July 1993 the Royal Astronomical Society of New Zealand held a joint meeting with the Astronomical Society of Australia. It was the first time that this had ever happened; the occasion marked the 70th Annual General Meeting of the RAS of New Zealand, and the 27th AGM of the Astronomical Society of Australia. The meeting was held in Christchurch, and was generally agreed to be a great success.

As one of the two Honorary Members of the RAS of New Zealand I had been invited to deliver the annual Harley Wood Lecture, and I duly did so. Subsequently it was suggested that I might produce a book about the stars of the far south. I was delighted to comply, but there is one point to be made at once. Though the book was planned in New Zealand, and was to be written for New Zealand readers, it is equally valid for Australia, South Africa, and those parts of South America where English is widely spoken — in fact, anywhere from Darwin through to Invercargill. There are differences in what can be seen; for example the brilliant northern star Capella can be seen from Auckland and the whole of mainland Australia, but not from the furthest part of New Zealand's South Island, while the even more brilliant Canopus is always above the horizon from Wellington, but sets over Sydney or Cape Town. However, these differences are minor, and in the text and in the maps I have allowed for them.

Astronomy, both professional and amateur, has a great following in all these lands; the skies are much clearer than those of Europe, and — rather unfairly, it is sometimes said! — some of the most interesting objects in the heavens are permanently out of view from European latitudes, so that today the main emphasis is on the southern hemisphere. I have tried to give a balanced view. Whether or not I have succeeded must be left to others to judge.

Part 1
The Sky Above Us

Chapter 1
Setting Out

New Zealand, Australia and South Africa are countries of many climates, many peoples and many moods. The Southern Alps are quite unlike the Nullarbor Plain or the South African veldt; Wellington, Perth and Johannesburg are very different cities. Yet one thing is common to them all: their night sky. Over much of the southern hemisphere the air is clear and calm, so that when darkness falls, the stars blaze down with an intensity unknown in foggier lands.

Everyone is familiar with the stars, and there can be few people who cannot recognize the Southern Cross; but how many children and how many adults have ever taken the trouble to find out what the stars are really like? Comparatively speaking, very few. It is often supposed that astronomy, the study of the sky, is a sort of mystic science, inaccessible to anyone who lacks a technical background and a huge, expensive telescope. Nothing could be further from the truth. Everyone can take an intelligent interest, and there is a great deal to be seen even with the naked eye, while a pair of binoculars will give you endless enjoyment.

What I aim to do, in this book, is to give an idea of what can be seen by anyone who is prepared to make the effort. I am not suggesting that you should make a task of it, or spend your evenings in intense concentration — only the select few will want to do that! But if you will bear with me for the next few pages, I hope that you will be interested enough to persevere, so that next time you find yourself under a starlit sky you will pause to look upward rather than shrug your shoulders and go indoors.

I began to take real notice of astronomy when I was a boy of six (candour compels me to admit that this was in 1929). I happened to pick up a book belonging to my mother, who was mildly interested in the subject, and I found it so fascinating that I followed it up. I firmly believe that the procedure I followed was sound, so I am quite prepared to pass it on. I recommend that you should:

1. Do some reading to make sure you know the basic facts.
2. Go outdoors at night, armed with a star map, and learn your way around the constellations. (It takes a surprisingly short time. I remember making a pious resolve to learn one new constellation every clear night, and within a few weeks I had a useful working knowledge of at least the main groups.)
3. Obtain a pair of binoculars, and start to look at some of the more interesting and spectacular objects.
4. Join an astronomical association. There are national societies in New Zealand, Australia and South Africa, and most large cities have groups of enthusiasts, often with observatories.
5. Then, if you are still interested, consider obtaining a telescope, always assuming there is not a convenient local observatory nearby which Society members can use.

Professional astronomy, of course, is quite another matter, and here a good science degree is essential; but everyone has to make a start, and it is worth remembering that astronomy is still one of the very few sciences in which the amateur can play a valuable role. Amateurs discover comets and novae; they make useful observations of the planets and of variable stars; they record polar lights and meteor showers, and much else. Today there are amateurs who make use of highly sophisticated equipment, but others use fairly basic instruments with tremendous effect. For example, one of the world's leading discoverers of those vast stellar explosions known as supernovae is an Australian country clergyman, the Rev. Robert Evans, who is equipped with a modest telescope which can even be moved around.

So let me begin with a 'roll-call' of the objects in the sky. Many of you will know these very elementary facts already. If so, you have my full permission to skip the next few pages and proceed at once to Chapter 3!

Chapter 2
The Universe Around Us

The Earth, our home in space, may be of supreme importance to us, but it is a very junior member of the Sun's family. It is a planet 12,756 km in diameter, moving round the Sun in a period of one year — or, to be more precise, 365.25 days. It is not the only planet; there are eight others, some larger than the Earth and some smaller. Most of the planets have secondary bodies or satellites moving round them. We have one satellite, our familiar Moon. Saturn, the ringed planet, has as many as eighteen known attendants, though admittedly most of them are relatively small and only one is larger than our Moon.

The distance between the Sun and the Earth is 150 million km. This sounds a long way, and in everyday life it is a very long way indeed, but it does not seem far to an astronomer, who is used to dealing with immense distances and immense spans of time. I do not pretend to be able to appreciate even one million km, let alone 150 million, but we know that these distances are correct, and we simply have to accept them.

There is nothing special about the Sun, except that we happen to live on a planet moving round it. The stars you can see on any clear night are themselves suns, many of them a great deal larger, hotter and more luminous than ours; for example the brilliant Canopus is thought to be the equal of about 200,000 Suns put together, while the red star Betelgeux, in Orion, is so large that its globe could swallow up the entire yearly path of the Earth.* The only reason why the stars seem so much less imposing is that they are so much further away. During the days when imperial measure was used, there was a very convenient 'scale model'; if you represented the Earth-Sun distance by one inch, the nearest star would be just over four miles away. Translate this into metric; if the Earth-Sun distance is represented by this line ------------------------, then the nearest star will lie at a distance of 6.7 km. No doubt many of these 'other suns' have planets of their own moving round them, and there is a good deal of indirect evidence that this is so, but as yet we have no cast-iron proof.

* The distances and luminosities of remote stars such as Canopus are bound to be uncertain to some extent, and different catalogues give rather different values. In this book I have followed the authoritative Cambridge catalogue, which seems to be as reliable as any.

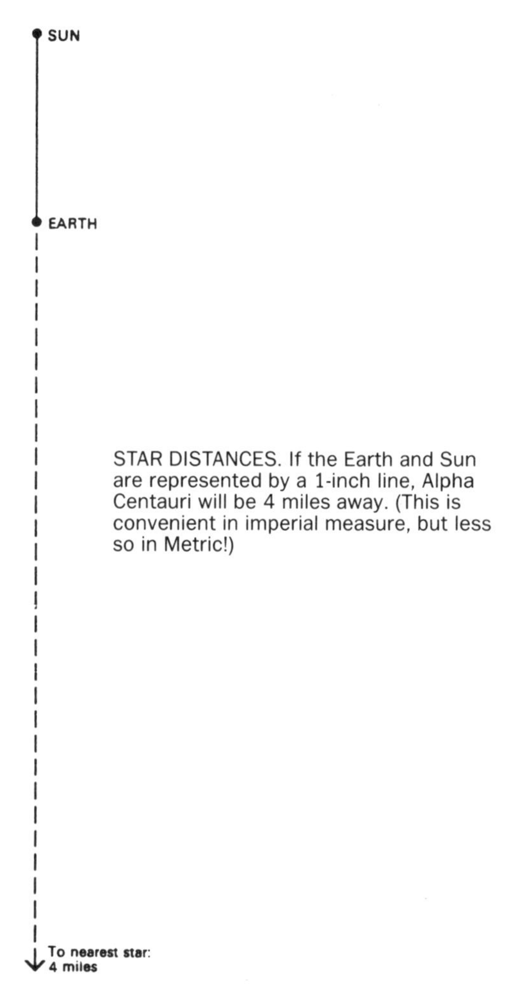

STAR DISTANCES. If the Earth and Sun are represented by a 1-inch line, Alpha Centauri will be 4 miles away. (This is convenient in imperial measure, but less so in Metric!)

The distances of the stars are so vast that ordinary units, such as the kilometre, become inconveniently short, just as it would be cumbersome to measure the distance between Auckland and Sydney in millimetres. Fortunately, Nature has provided us with an alternative. Light does not travel instantaneously; it flashes along at a speed of 300,000 km per second. In a year, therefore it covers 9.46 million million km, and this is known as the 'light-year', which, please note, is a measure of distance, not of time. The nearest bright star beyond the Sun, Alpha Centauri (the brighter of the two Pointers to the Southern Cross) is 4.2 light-years away, corresponding to around 40 million million km.

Early men divided up the stars into patterns or constellations. The first star maps were due to the great civilizations of antiquity, notably the Chinese, the Egyptians and the Chaldaeans; the system we use today is basically Greek, though it has been altered and extended since those far-off days. Many of the constellations have mythological legends associated with them. For example, many people will have heard the legend of Perseus and Andromeda — how the beautiful Princess Andromeda was chained to a rock by the sea-shore to be devoured by a monster, because her mother, Queen Cassiopeia, had been unwise enough to offend the sea-god Neptune by boasting that her daughter was more beautiful than the sea-nymphs; how Cepheus, king of the stricken realm, was forced to offer up Andromeda as a sacrifice, and how the princess was rescued by the dauntless hero Perseus at something later than the eleventh hour. All the characters in the legend are to be found in the sky, though the monster is often relegated to the status of a harmless whale. Other constellations have more modern names, because they could not be seen from northern latitudes; thus we have an Air-Pump, a Telescope, a Microscope and an Octant.

Yet the constellation patterns actually mean nothing at all, because the stars are at very different distances from us, and we are dealing with nothing more significant than line of sight effects. To show what is meant, consider the Southern Cross, where we have four bright stars making up a sort of kite pattern (it is a pity that there is no central star to produce a true X). These four are known to astronomers as Alpha, Beta, Gamma and Delta Crucis. Their distances from us, in light-years, are respectively 360, 125, 88 and 257, so that, for example, Alpha is much further away from Gamma than we are. If we were observing from a different vantage point, the four stars might well be in completely different parts of the sky.

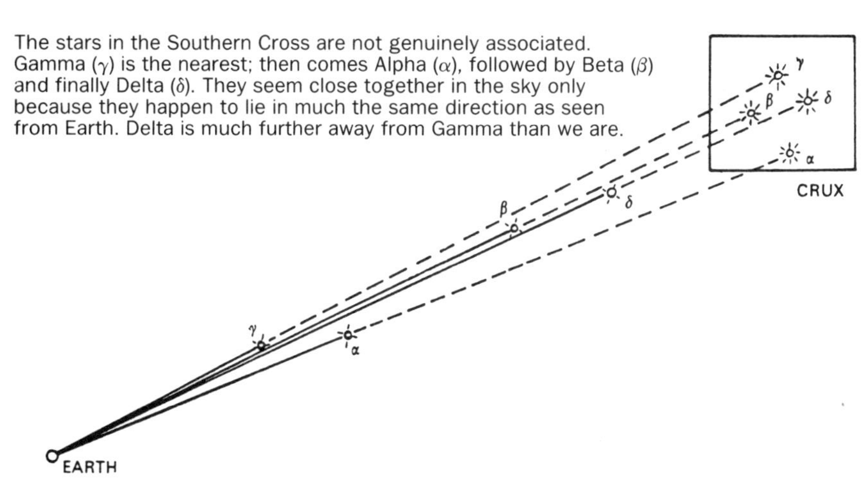

The stars in the Southern Cross are not genuinely associated. Gamma (γ) is the nearest; then comes Alpha (α), followed by Beta (β) and finally Delta (δ). They seem close together in the sky only because they happen to lie in much the same direction as seen from Earth. Delta is much further away from Gamma than we are.

(*En passant*, this may be the place to dispose of the pseudo-science of astrology, which claims to link the positions of the stars and planets with human character and destiny. Of course, astrology was understandable enough in ancient times, but it has no basis whatsoever, and there is no excuse for it today. When asked for my views about astrology, I usually say that it proves only one scientific fact: 'There's one born every minute'! Astrology still has its devotees — but, of course, there are also some people who continue to believe implicitly in Father Christmas.)

The Sun is a member of the star-system we call the Galaxy, which is made up of roughly 100,000 million stars. The Galaxy is a flattened system, shaped rather like a double-convex lens, though I personally prefer the less scientific description of it as resembling the shape of two fried eggs clapped together back to back. The overall diameter is about 100,000 light-years. The Sun, with its family or Solar System, lies near the main plane, about 30,000 light-years from the galactic centre. When we look along the main plane, we see many stars in almost the same direction, and this produces the effect of the Milky Way, that lovely band of light which can be seen stretching across the heavens from one horizon to the other. Binoculars are enough to show that the Milky Way is made up of stars, and it may seem that these stars are in danger of bumping into each other — but again appearances are misleading; the stars in the Milky Way are not genuinely crowded together, and stellar collisions must be very rare indeed.

We cannot see through to the centre of the Galaxy, because there is too much obscuring 'dust' spread between the stars, but we do know that the centre lies beyond the glorious star-clouds in the constellation of Sagittarius, the Archer. We also know that the Galaxy is rotating. It takes the Sun about 225 million years to make one full revolution round the nucleus, and this is the period often referred to as the cosmic year. One cosmic year ago, the most advanced life-forms on Earth were amphibians; even the ferocious dinosaurs lay in the future. It is interesting to speculate about what conditions will be like here one cosmic year hence.

Our Galaxy is not the only one. Far away in space we can see others, most of them so remote that their light takes millions, hundreds of millions or even thousands of millions of years to reach us. Indeed, the only two galaxies which are close enough to be prominent naked-eye objects are the two Clouds of Magellan, about which I will have much more to say later; but even the Large Cloud of Magellan is 169,000 light-years away, so that we are seeing it as it used to be 169,000 years ago. Once we go beyond the Solar System, our view of the universe is bound to be very out of date.

Even the Clouds must be regarded as our next-door neighbours. The most remote systems known to us are at least 13,000 million light-years away, so that we are seeing them as they used to be long before the Earth existed; the age of the Earth is known, with fair accuracy, to be about 4600 million years.** We have also found out that the entire universe is expanding, so that all the galaxies beyond our own local group of systems are racing away from us — and the further away they are, the faster they are going; the most distant galaxies are receding at well over 90 per cent of the velocity of light. This does not mean that we are particularly unpopular, because every group of galaxies is receding from every other group. There is nothing at all special about our own position in the universe.

How did the universe begin — and how will it end, if indeed it will end at all? And how large is the universe? These are fundamental questions, and as yet we cannot pretend to answer them; there is a great deal that we do not know, and which we may never know. I will return to these problems later, though I cannot hope to say anything really constructive. Meanwhile, let us come back to our own particular area.

The Solar System is made up of one star (the Sun); nine known planets, and various bodies of lesser importance, such as the moons or satellites; the flimsy and ethereal comets, and assorted space debris. The planets are divided into two well-defined groups. First we have four small, solid worlds: Mercury, Venus, the Earth and Mars. Then comes a wide gap, in which move thousands of midget worlds known as minor planets or asteroids. Beyond come the four giants, Jupiter, Saturn, Uranus and Neptune, which seem to have solid cores but are made up chiefly of liquid, with gaseous surfaces. There is also a strange little interloper, Pluto, which is very much of a maverick and is probably unworthy of true planetary status. Neptune, outermost member of the main system, is on average 5900 million km from the Sun, and takes almost 165 years to make one circuit; by contrast the innermost planet, Mercury, has a mean distance from the Sun of a mere 60 million km, and has a 'year' equal to no more than 88 Earth-days.

The Sun, like all normal stars, is made up of hot gas, and is creating its own energy — not by simple burning in the manner of a coal fire, but by nuclear transformations taking place deep inside it. It is big enough to engulf over a million bodies the volume of the Earth, and it is the absolute ruler of

** Many people call this '4.6 billion'. But the English billion is a million million, while the American billion is a thousand million; and since this can cause confusion, I prefer to avoid the term 'billion' altogether.

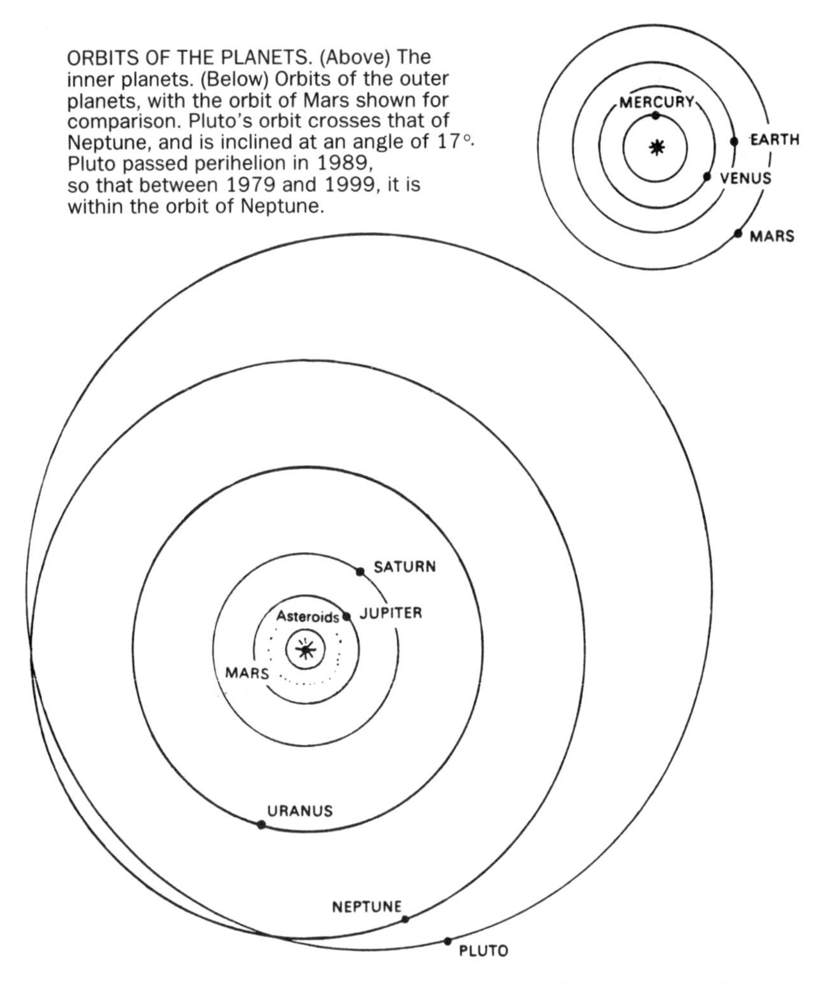

ORBITS OF THE PLANETS. (Above) The inner planets. (Below) Orbits of the outer planets, with the orbit of Mars shown for comparison. Pluto's orbit crosses that of Neptune, and is inclined at an angle of 17°. Pluto passed perihelion in 1989, so that between 1979 and 1999, it is within the orbit of Neptune.

the Solar System. Even Jupiter, by far the largest and most massive planet, has a diameter of only 143,884 km, as against 1,392,00 km for the Sun.

The planets have no light of their own; they shine only because they reflect the rays of the Sun, in the manner of large, though (in most cases) not very efficient, mirrors. Therefore they look like stars, and some of them are brilliant; the first five have been known since ancient times, though the outer three have been discovered only more recently — Uranus in 1781, Neptune in 1846 and Pluto in 1930. Uranus can just be seen with the naked eye if you know where to look for it, but to see Neptune you will have to use binoculars, and Pluto is invisible without a telescope of fair size.

The planets betray their nature by the fact that they move slowly around from one constellation to another. The stars themselves are not genuinely fixed in space, and are moving around in all sorts of directions at all sorts of speeds, but they are so far away that their individual or 'proper' motions are too slight to be noticed over periods of many lifetimes. The patterns seen by the early civilizations are to all intents and purposes the same as those of today, and the first Maori or Aborigine star-gazers too will have been able to see the Southern Cross and all the other groups which we know so well. The planets, which are so much closer to us, shift very obviously from one night to another, and indeed the word 'planet' really means 'wanderer'.

The Moon, like the planets, depends entirely upon reflected sunlight. It seems so glorious to us only because it is so near; its mean distance from us is only 376,284 km — and this is less than twenty times the distance between, say, Auckland and London. It is moving round the Earth in a period of just over 27 days, and it shows regular phases or changes of shape, simply because the Sun can light up only half of it at any one time, and everything depends upon how much of the sunlit side is turned in our direction. The Moon has no air, no oceans and no life; the dark markings seen on its surface are known as seas, but there has never been any water in them.

Among the less important but sometimes very spectacular members of the Solar System are the comets. Unlike the planets, comets are not ponderous and massive, and the nucleus of even a major comet, consisting of ice mixed with 'rubble', is no more than a few kilometres in diameter. Most comets move round the Sun in paths or orbits which are not almost circular, as those of the main planets are, but are very elliptical. When a comet nears the Sun, the ices in its nucleus start to evaporate, so that the comet develops a 'head' and sometimes a long tail or tails. Early men were terrified of comets, but the chances of our being hit by one are very low, and even if it happened, it would not mean the end of the world, as was sometimes suggested.

Meteors are cometary debris, and are tiny particles, usually smaller than grains of sand. They move round the Sun in the same way as we do, but unless they dash into the Earth's air we cannot see them. If they do come into the upper atmosphere, moving at anything up to 72 km per second, they set up so much friction against the air-particles that they burn away in the streaks of luminosity which we call shooting-stars. Clearly there is no connection between a real star, which is a sun, and a tiny, friable meteor!

Note also that while a meteor flashes across the sky, a comet does not. If you see an object which is moving obviously against the starry background, it cannot be a comet. If the motion is very rapid, it is probably a meteor; if it is slow, then it is more likely that you are seeing an artificial satellite.

There are two ways of exploring the sky. One is by observation, and this has been done ever since the dawn of human history, though telescopes have come upon the scene only during the past few hundred years. The other is by space-craft. The Space Age began on 4 October 1957, when the Russians sent up their first artificial satellite or man-made moon, Sputnik 1, which sped round the world and transmitted the 'Bleep! bleep!' radio signals which will never be forgotten by anybody who heard them (as I did). Since then, space-craft have been sent past all the planets except Pluto; men have landed on the Moon, and unmanned vehicles have touched down upon the surfaces of Venus and Mars.

A journey to the Moon takes only a few days, and Neptune, the outermost giant planet, was by-passed in 1989 by a space-craft which had been launched in 1977. But to send a probe to even the nearest star would take so many centuries that for the time being we can discount it. By our present-day standards, interstellar travel, as opposed to interplanetary travel, is out of the question. This is not to say that it will always be impossible. After all, television would have been sheer science fiction when the first white men arrived in New Zealand, but we must await some fundamental 'breakthrough', about which we cannot even speculate as yet because we have nothing to guide us.

Such are the main members of the great universe in which we live. Whether we are alone, or whether there are other civilizations, we cannot tell; it seems logical to assume that there is life elsewhere, but up to the present time we have no proof, though we are doing our best to find out. We have at least come a long way since early men first gazed upward and marvelled at the starlit sky, though we have to admit that we have still a great deal to learn.

Chapter 3
Star-Gazing Through The Ages

Early men believed the world to be flat, and to lie at rest in the exact centre of the universe. Both these ideas were quite logical at the time. After all, the Earth really does look flat, allowing for local variations such as hills and valleys, and the entire sky seems to revolve around it once a day. It was equally reasonable to assume that the stars were tiny lights attached to a solid, invisible sphere, and that the moving planets were much closer to us.

One by one these mistakes were discovered. The Greeks made scientific observations to show that the Earth cannot be a flat plane; for example the star Canopus can be seen from Alexandria but not from Athens, which would be inexplicable unless the Earth is a globe. One of the Greek philosophers, Eratosthenes, even made a successful effort to measure the circumference of the Earth — and his value was much more accurate than that used by Christopher Columbus in his epic voyage undertaken in 1492. Had Columbus adopted Eratosthenes' figure, he would have realized that he could not possibly have reached India, as he fondly believed.

The second error, that of the Earth's position in the cosmos, persisted for much longer. Aristarchus of Samos, one of the most brilliant of the old Greeks, did relegate the Earth to the status of a planet orbiting the Sun, but he could give no conclusive proof, and he found few followers. The great revolution in outlook was not really complete until the publication of Isaac Newton's great book, the *Principia,* in 1687.

Of course, all the early star-gazers whose work has come down to us were northern-hemisphere dwellers. Southern astronomy, in the true sense, did not begin until much later, but obviously the early peoples of the south did take a lively interest in the sky, and they had their own folklore and their own legends — the Maori of New Zealand, the Aborigines of Australia and the various African tribes. Unfortunately we do not know as much about these old tales as we would like, because they were not particularly well documented at the time and now it is too late.

All nations have their own creation myths. To the Maori it seems that the supreme being was Tangotango, who mated with Moeahuru to produce the celestial bodies. He was the first-born of the Sky Father and Earth Mother, and he is responsible for the alternation of day and night.

One Maori tale tells how the Sun was 'tamed'. Initially, it is said, the Sun moved across the sky much more quickly than it does now, and was so hot that the ground below was scorched. Eventually Maui, one of the great characters of Maori lore, managed to catch the Sun in a net as it rose above the horizon, and belaboured it until it promised to slow down and produce less heat. There is also a delightful tale concerning the Maori equivalent of Europe's 'Man in the Moon', supposedly seen in outline because of the chance arrangement of the bright and darker areas of the lunar surface. In Maori lore the 'Man in the Moon' becomes a woman, Rona, who was visiting a spring one moonlit night in order to collect water in her gourd. When the Moon passed behind a cloud, Rona called it an offensive name. At once she was snatched up and placed on the Moon, where she can still be seen, together with her rururu taha or bundle of gourd-vessels.

The Maori had their own star-groups one of which was Te Ra i Tainui or the Sail of Tainui, a canoe; the star cluster of the Pleiades forms the bow, while the three stars we now call Orion's Belt represent the stern. The sail itself is made up by the second famous star-cluster, the Hyades, while the cable is marked by the two Pointers to the Southern Cross, and the Cross itself is the anchor (Punga a Tama-rereti). The Pleiades cluster is conspicuous as seen with the naked eye; keen-sighted people can see at least seven individual stars in it, and it is claimed that the Maori could see many more. Its first appearance in the dawn sky each spring marked the start of the Maori year, and there were many celebrations, including displays of dancing. Other important stars were Canopus (Aotahi), whose first appearance in the dawn sky each year was the signal to start planting certain crops; Vega (Whanui), which announced that the harvest was imminent; and Rigel in Orion (Puanga). In the Chatham Islands it was the first dawn appearance of Rigel, rather than the Pleiades, which was used to mark the start of the new year. The Pleiades themselves represent a group of girls playing for the benefit of youths represented by the stars of Orion, while the orange star Aldebaran is an old man acting as timekeeper for the dancers, and the Milky Way is a river in which the sky people catch fish for food.

Meteors were regarded as unlucky, and eclipses of the Sun, now known to be due to the temporary covering-up of the Sun by the dark body of the Moon, were attributed to attacking demons — the equivalent of the dragons of the ancient Chinese. There was also a god, Maru, apparently associated with the lovely glows known as aurorae or polar lights, quite often seen from New Zealand (particularly from the South Island). Aurorae are due to the electrified particles from the Sun, which cascade into the upper air and produce glows. To the Maori they could be either lucky or unlucky according to their form; if seen in the form of a bow behind a travelling war party, it foretold victory, but if the aurora were incomplete it was regarded as a bad sign.

In Australia, the Aborigines believed that originally the Earth was lifeless; then, during what has been called the Dreamtime, came giant semi-human beings who eventually produced the world of today. Some of the bright stars attracted special attention. I have heard that the appearance of the northern orange Arcturus was a sign that it was time to set fish traps. In one part of the country it is said that the Southern Cross represents a sting-ray being chased by a shark, while elsewhere the Cross is a white gum tree and the Pointers are two yellow-crested cockatoos trying to roost in the tree.

One of the stories of the Sun is remarkable. It is said that there was an argument between an emu and a brolga about which of their chicks was the more beautiful. After an acrimonious exchange of views, the brolga ran to her rival's nest and threw the emu egg into the sky, where it shattered against sticks collected by the sky people. The yolk of the egg burst into flame and became the Sun, and the sky people, amazed by the beauty of the Earth as illuminated by the fire, decided that in future the Earth people should be given night and day.

African legends are, predictably, very varied, and again are not well documented, but the record is not entirely blank. Since very early times the African tribes have known that the Earth is in motion relative to the Sun, and there is an old song which, when translated, means 'I shall worship you and go round you as the Earth worships the Sun'. Cave engravings show the five naked-eye planets revolving round the Sun. Venus was of special significance, because it was believed that any child born early in the morning when Venus was visible would be beautiful — and would have many lovers. One such child was Nandi, mother of the famous Zulu leader Shaka. Mars, of course, was thought to have an evil influence; nothing seems to be on record about Mercury, Jupiter or Saturn.

In African mythology the Earth was a bowl made of clay and rocks, criss-crossed by mountains and rivers; it was carried on the back of a giant tortoise, which floated around the Sun in a vast, blue ocean called the Sea of Eternity. It was believed that when the tortoise tips to one side, the blue surface can be seen as well as the Sun, and the solar rays shine into the bowl, causing daylight; when the tortoise floats upright, the edge of the bowl acts as a shield against the Sun, revealing nothing but the blackness of Heaven.

From this, it seems that the Africans had at least some kind of cosmogony, and presumably there was a great deal of sun-worship and moon-worship. To the Pygmies the supreme god was Khonvum, who controls all the phenomena of the sky. When the Sun dies at nightfall, Khonvum collects the broken pieces of stars in his sack and throws armfuls of them at the Sun, mending it so that it can rise unharmed next morning. In a legend from the Zambezi, the Moon 'used to be very pale, and did not shine; she was jealous of the glittering Sun. Once, when the Sun was looking at the other side of the Earth, the Moon stole some of his fire to adorn herself. In anger, the Sun splashed the Moon with mud, and the dark patches can still be seen. But the Moon was bent on vengeance, and whenever she can catch the Sun off his guard she spatters him with mud. Then the Sun stops shining for some hours, and the whole Earth is in terror'. Is this a reference to solar eclipses?

The Moon was important in Zulu lore because it was thought to influence animals and plants. For instance, it was held that peas planted between two and three days before full moon yielded more than twice the crop of those planted earlier. No woman 'at the time of her Moon' was allowed to eat amasi (curdled milk) or to become angry — or to enter a cattle-kraal, or to sit on the wooden stools normally used by the men. Among the Bushmen, the lunar phases were explained quite simply; the Moon regularly offends the Sun, and is whittled away before crying for mercy and being gradually restored. The Bushmen known as the Ba-letsatsi, or Men of the Sun, were disinclined to work on a day when the Sun rose behind clouds, on the pretext that it might damage their hearts. It has also been said that the Bushmen held their ceremonial dances at the times of new moon and full moon. Dancing began with the new moon, because the dark nights have ended; it was continued at full moon, so that those taking part could enjoy the delightful coolness after the heat of the day.

In some African paintings the Sun is shown as a circle with a cross through it — in fact, a swastika, the oldest symbol of light in Africa. The full moon is a circle with a central dot, indicating Love; the crescent moon indicates Tragedy. Moreover, there are records of the Three Healers (Orion's Belt), the Ploughing Stars (the Pleiades), the Starry River (the Milky Way) and the Tree of Life (the Southern Cross). The Cross is the holiest of all symbols, put into the sky specifically to act as a guide to travellers who would otherwise lose their way.

All this is fascinating, but it is not science in the Western sense of the word, and inevitably southern hemisphere astronomy in the true sense did not begin until the arrival of Europeans: South Africa came first, then Australia, and finally New Zealand.

The South African story began with the arrival of a redoubtable Jesuit priest, Father Guy Tachard, who came ashore at the Cape in 1685 when he was on his way to Siam, and stayed long enough to establish a temporary observatory, mainly to make some measurements to help in determining the precise longitude of the Cape. At this time the Dutch governor of the Cape was one Simon van der Stel, who gave Tachard's party a warm welcome and did all he could to make them comfortable. Tachard was equally anxious to be courteous, and at the outset a curious muddle occurred with regard to firing the correct number of gun salutes. To quote Tachard's own words:

'It was agreed upon that the fort should render Gun for Gun when our Ship saluted it. This Article was ill explained, or ill understood, by these Gentlemen, for about ten of the Clock my Lord Ambassador having ordered seven Guns to be fired, the Admiral answered with only five Guns, and the Fort fired none at all. Immediately the Ambassador sent ashore again, and it was determined that the Admiral's Salute should pass for nothing, and so the Fort fired seven Guns, the Admiral seven Guns, and the other Ships five, to salute the King's Ship, which returned them their Salutes, for which the Fort and Ships gave their Thanks.'

All was well, and Tachard stayed at the Cape for several pleasant weeks, finally departing with a gift from his hosts of 'Tea and Canary Wine'. The next astronomer to visit the Cape, Peter Kolbe, was not a success — apparently he was anything but a teetotaller! But then, in 1751, came the Abbe de La Caille, a renowned French astronomer whose aim was to draw up a catalogue of the far-southern stars. The only previous attempt had been made by Edmond Halley, who is probably best remembered today because of his connection with the famous comet — about which I will have more to say later. Halley had visited the island of St Helena in 1677 and had done some useful work, but it was preliminary only, and La Caille was much more ambitious. He stayed in South Africa for less than two years, but during this time he catalogued 10,000 stars.

The first official observatory in South Africa was set up at the Cape in 1821, with an Englishman, the Rev. Fearon Fallows, as Director. Fallows did not have an easy time, and the difficulties he faced probably contributed to his early death in 1831. His successor, Thomas Henderson, stayed for only a brief period because he detested the place — he referred to it as a 'dismal swamp' — but at least he made the observations which led to the measurement of the distance of a star, Alpha Centauri, which Henderson correctly found to be just over four light-years away. Next came Thomas Maclear, who really established the Cape Observatory as a leading institution. During his regime, too, there was a visit from Sir John Herschel, one of the world's leading astronomers, who stayed from 1834 to 1838, and made the first really exhaustive survey of the southern sky.

Since then the Cape Observatory has maintained its position of eminence, and during the twentieth century other observatories also flourished, such as those at Johannesburg, at Boyden near Bloemfontein, and for a while the Radcliffe Observatory near Pretoria. The last of these was abandoned in 1974, because Pretoria had grown alarmingly, and was not only larger (which did not greatly matter) but also brighter (which mattered very much; light pollution is the scourge of the present-day astronomer). The South African authorities made the courageous decision to move most of their main equipment to a new site at Sutherland, in Cape Province, where conditions are excellent. The main telescope at Sutherland is now a 188-cm reflector, shifted there from the old Radcliffe.* Boyden remains active, with the 152-cm Rockefeller telescope as its main instrument, and the old 62-cm refractor remains at Johannesburg, at the Republic Observatory.

Australian astronomy began during the eighteenth century. Everyone knows that Captain Cook first sighted the Australian coast in April 1770, but not everyone knows that the prime purpose of his voyage had been to take astronomers to observe the 1769 transit of Venus. Venus, remember, is closer to the Sun than we are, so that it can sometimes pass in front of the solar disk, appearing as a small black circle. This does not happen often — the last occasion was in 1882, the next will be in 2004 — and transits used to be regarded as very valuable, because they provided an opportunity to make measurements of a kind which would lead to a better estimate of the distance between the Earth and the Sun. As the whole method is now completely obsolete there is no point in saying more about it here, but in any case Cook was detailed to take Nathaniel Green and other astronomers to the South Seas to observe the transit of 1769 — a 'once in a lifetime' mission, because the next transit was not due until 1874. Cook reached Tahiti, and the observations were successful. Only then did Cook open the sealed orders to find

* There is a delightfully comic episode connected with the Radcliffe. When the Observatory was established, local roads were named after stars: Rigel Road, Arcturus Road, Capella Road and so on. One name allotted was *Canopsus* Road. It was pointed out to the Pretoria City Council that this was a mistake; it ought to be Canopus. Their reply has gone down in history: 'Our maps have been printed and distributed, and cannot be altered now. Can you not alter the name of the star?'

that his next task was to search for the expected Southern Continent.

The first Australian observatory followed in 1786, when a Naval officer, Lieutenant Dawes, arrived with instructions to make measurements to help in longitude determination. Dawes stayed only until 1791, but then a new Governor, Sir Thomas Brisbane, was appointed. He was an enthusiastic amateur astronomer, and it was even claimed that he spent more time upon astronomy than upon his official duties. This is probably unfair, but it undoubtedly was a factor in his premature recall. Brisbane set up a fully-fledged observatory behind Government House in Parramatta, and much valuable work was carried out there; observations of comets, star-clusters and nebulae were undertaken by Brisbane's chief assistants, Rumker and Dunlop, and the 'Brisbane Catalogue of Stars' had 40,000 entries. Sadly, the Observatory no longer exists. It was closed in 1847, and only an obelisk now marks its site.

However, other major observatories followed at Melbourne, Sydney, Adelaide, Perth, Brisbane and Hobart. Some of them had rather chequered careers — particularly Melbourne, where a large telescope was set up but proved to be a failure. Only in modern times has it been re-built and put to good use (it is now in the Mount Stromlo Observatory in Canberra). Yet today, Australian astronomy is very much in the ascendant. The main observatories are those of Mount Stromlo and at Siding Spring, near Coonabarabran, where the AAT or Anglo-Australian Telescope, with its 3.9-m mirror, is one of the finest in the world, and is fully computerized. Also at Coonabarabran is the UKS or United Kingdom Schmidt Telescope, capable of covering wide areas of the sky with a single photographic exposure.

Then, too, there is the radio telescope at Parkes.

Light is a wave motion, and the colour of the light depends upon its wavelength; for visible light, red has the longest wavelength, violet the shortest. If the wavelength lies outside these limits, the radiations do not affect our eyes, though they can be detected in other ways, To the short-wave end of violet light we have ultra-violet, X-rays and finally gamma-rays; to the long-wave end of red light we have infra-red, micro-waves and then radio waves. All these make up the total range of wavelengths, or electromagnetic spectrum.

Many of the radiations are blocked out by the Earth's air — for example, virtually no X-rays can pass through — but some radio waves can penetrate the atmosphere, and are collected by instruments known as radio telescopes. The name is misleading, because a radio telescope is not in the least like a visual telescope; and one certainly cannot look through it; the end product is usually a trace on a graph. The Parkes telescope, completed in 1961, collects the radio waves by means of a 64-m 'dish'. Though it is now over 30 years old, it remains one of the largest and most effective of all radio telescopes, and it has been responsible for many fundamental advances. By now there is a whole network of radio telescopes spread over Australia, working together — for example the installations at Tidbinbilla — and these can give us information which we could never obtain in any other way.

Amateur astronomy in Australia has always been very strong. One of the leaders was John Tebbutt, who set up an observatory at Windsor. He discovered two brilliant comets — those of 1861 and 1881 — and made thousands of observations which his professional colleagues found very useful indeed. There are many present-day amateur astronomical societies in Australia, including the New South Wales Branch of the British Astronomical Association of which Tebbutt was an early member. The tradition now is as strong as ever; I have already mentioned the Rev. Robert Evans, the supernova-hunter, who has already found more than two dozen of these remarkable exploding stars in remote galaxies.

New Zealand astronomy began rather later than that of Australia, but observations were already being made during the second half of the last century. Go to Queenstown, and you will find a tablet marking the site where astronomers went to observe the 1874 transit of Venus. An observatory for navigational research was established in 1868, and amateur astronomy flourished; for example there was John Grigg, a celebrated comet-hunter. In 1896 Charles Rooking Carter left a large sum of money to start a fund for an observatory to be based in Wellington. The project took a long time, but the Carter Observatory was finally opened, in 1941, with Dr. Murray Geddes as Director. The main telescope is an old but good 9-inch (23-cm) refractor.

In 1977 the Carter became recognized as New Zealand's national observatory, but it has to be admitted that observing conditions there are poor; Wellington is too bright. The main observational centre is now at Mount John, overlooking Lake Tekapo in the plains of the Mackenzie Basin. The main telescopes are a 1-m reflector and two 61-cm instruments; they may be small by international standards, but they do have the advantage of latitude, and indeed Mount John is the southernmost major observatory in the world. There is an outstation at Black Birch in the Southern Alps, a site recommended by yet another of New Zealand's brilliant amateurs, Frank Bateson. There is an excellent observatory on One Tree Hill in Auckland. Amateur-run, it produces work of full professional standard, and is equipped with a 51-cm reflector.

A New Zealand Astronomy Centre has been established at the Carter Observatory, under the direction of Dr Wayne Orchiston, and it is very active in astronomical education; it has a small planetarium, and a larger planetarium is being planned for Auckland. The outlook should be bright, but there are clouds on the horizon. Money is in short supply, and the usual practice of governments under such circumstances is to close institutions such as hospitals and research centres, including observatories, rather than cutting back on armaments and futile contributions to warring countries overseas. As I write these words (April 1994) there is a real threat not only to Black Birch, but to the research side of the Carter Observatory itself. Let us hope that wiser counsels will prevail.

In any case, it is clear that southern-hemisphere astronomy is in a flourishing state, despite the constant financial problems. I appreciate that the account I have given here has been watered down to the point of dehydration, but I hope that it is enough to give a general idea of the situation — so let us now turn to more general topics, and consider what has come to be known as the celestial sphere.

Chapter 4
The Revolving Heavens

Men of many centuries ago used to believe in a flat Earth, mounted upon pillars rather in the manner of a table-top. When the Sun sank below the western horizon, at dusk, it spent the next hours threading its way through the pillars, miraculously avoiding touching any of them, and making its reappearance in the east at dawn the next day. Nowadays, we are rather less naive. We know that the east-to-west rotation of the sky is due not to any movement of the heavens, but to the real rotation of the Earth, which is spinning on its axis from west to east.

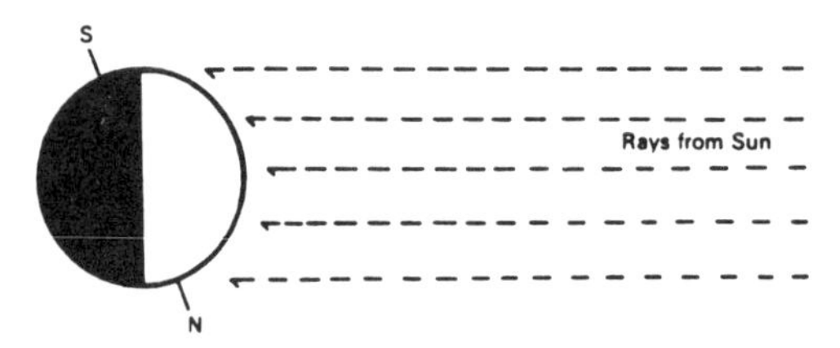

DAY AND NIGHT. The Sun can light up only half of the Earth at any one time.

Cause of the seasons.

There can be few people who do not know the cause of day and night, and of the rotation of what we call the celestial sphere. Just in case of any confusion, it may be helpful to include a diagram to show just what happens. As we have noted, the Earth moves round the Sun at a mean distance of 150 million km, taking 365¼ days or one year to make a full circuit. For the sake of convenience, we round off our calendar to 365 days, and add an extra day once every four years to allow for the extra quarter-day needed for a complete revolution.*

The Earth is spinning round in a period of 23 hours 56 minutes, rounded off in everyday life to 24 hours. The axis of rotation passes through both poles and also through the centre

* *En passant*, what will be the first day of the new century? Answer — 1 January 2001, not 1 January 2000, because there was no year 0. Rather illogically, the calendar goes straight from BC 1 to 1 AD.

of the Earth; in the diagram it is lettered SN, S representing the south pole and N the north pole. Obviously the Sun can light up only half of the world at once, so that day and night follow each other in regular succession.

However, day and night conditions are not the same everywhere on the Earth, because the axis of rotation is tipped or inclined to the plane of the orbit. (Orbit, let me add, is merely the astronomer's term for 'path'.) The angle is 23½° to the perpendicular, as shown in the next diagram. This is the same as saying that the angle between the orbital plane and the Earth's equator is 23½°. And this tilt of the axis is the cause of the seasons.

Let us begin in the position marked 'January'. The south pole of the Earth (S) is tilted toward the Sun, and so the southern hemisphere of the world is enjoying its summer; the polar regions are in constant daylight, and there is no darkness at all. Meanwhile, the northern hemisphere is experiencing its winter, and at the north pole there is no daylight, because the Sun remains below the horizon for the complete 24-hour rotation. In the position marked 'June', conditions are reversed, so that it is winter in the south and summer in the north. If the Earth's axis were 'upright', we would have no true seasons. This is virtually the situation on the giant planet Jupiter, where the year is 11¾ times as long as ours and the tilt of the axis is a mere 3°. (Not that this makes a great deal of difference, because Jupiter is five times as far away from the Sun as we are, and the temperature of the outer clouds is always so low that the absence of any seasonal effects is of no practical importance.)

There is a slight additional complication due to the fact that the Earth's orbit round the Sun is not a circle, but an ellipse. We are 152 million km from the Sun in winter (aphelion) and only 147,100,000 km away in summer (perihelion). The effects of this changing distance are slight, and are in any case masked by the consequences of the greater amount of ocean in the southern hemisphere, but they can be detected. The Earth reaches its nearest point to the Sun in late December. Although it is then high summer in New Zealand and Australia, it is midwinter in Europe, so that Londoners are having to cope with snow-drifts and sub-zero temperatures.

Because the Earth goes round the Sun once a year, the Sun seems to move right round the sky once a year; its apparent path against the starry background is known as the *ecliptic*. Of course, it is impossible to see the Sun and the stars at the same time, simply because the feeble starlight is drowned by the brightness of the sky, but calculations can always show just where the Sun is at any particular moment. During southern summer, the Sun is in the south part of the sky, and remains above the horizon for more than 12 hours out of the 24; in winter the Sun stays in the northern part of the sky. The ecliptic passes through the twelve constellations which make up what we call the Zodiac.

The celestial sphere, shown in the next diagram, is not nearly so complicated as it looks. For this purpose it will be convenient to go back to the ideas of the ancient Greeks, and assume that the sky really is a solid sphere, turning round the world once in 24 hours and taking the Sun, Moon, stars and all other celestial bodies with it. The celestial sphere is concentric with the surface of the Earth, so that it has the same centre (C in the diagram).

The Earth is divided in two by the equator (EQ). Similarly, the celestial sphere is divided in two by the celestial equator

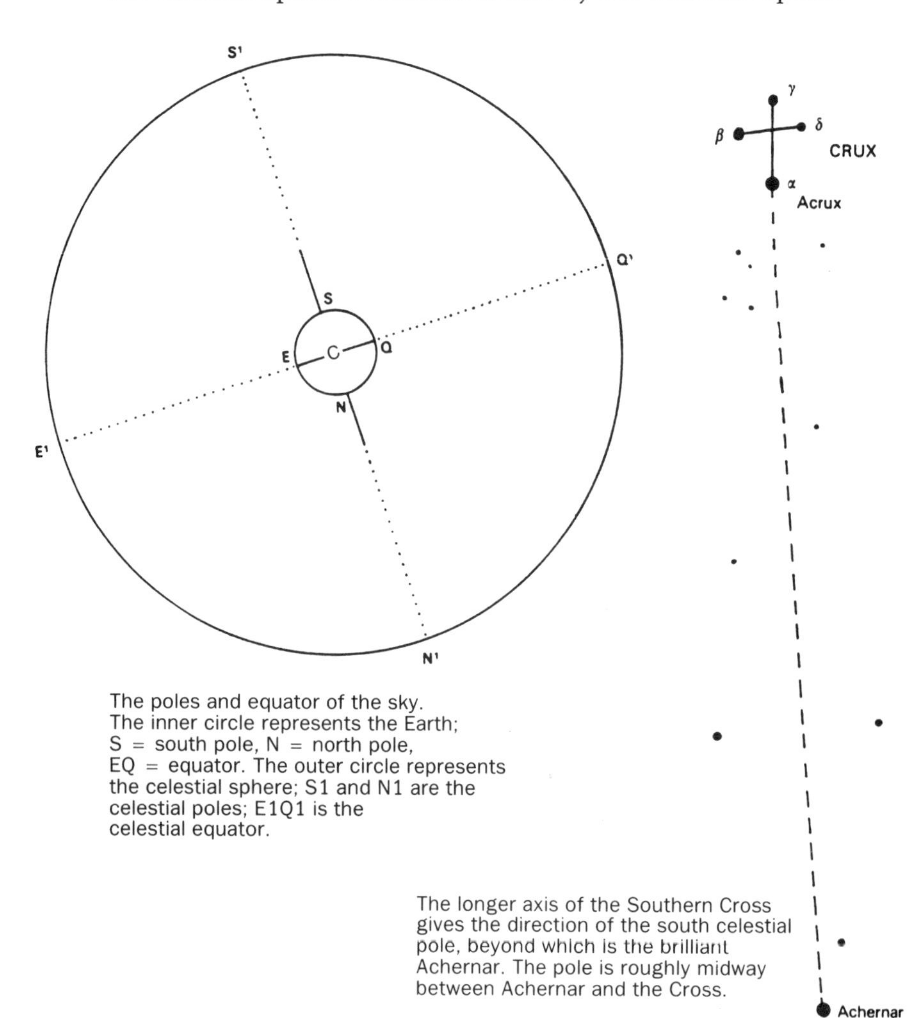

The poles and equator of the sky. The inner circle represents the Earth; S = south pole, N = north pole, EQ = equator. The outer circle represents the celestial sphere; S1 and N1 are the celestial poles; E1Q1 is the celestial equator.

The longer axis of the Southern Cross gives the direction of the south celestial pole, beyond which is the brilliant Achernar. The pole is roughly midway between Achernar and the Cross.

(E′Q′), so that there is a northern hemisphere and a southern hemisphere. (Put in more technical terms, the celestial equator is the projection of the Earth's equator on to the celestial sphere.) Next, let us continue the line of the Earth's axis until it hits the celestial sphere. In the diagram, S and N represent the two poles of the Earth, so that S′ and N′ represent the two poles of the sky.

The north celestial pole is conveniently marked by a bright star, Polaris in the constellation of Ursa Minor (the Little Bear), which lies within 1° of the polar point. Of course, Polaris can never be seen from anywhere south of the Earth's equator, and unfortunately there is no bright south polar star; the pole lies in a barren region, made up of the faint constellation of Octans (the Octant). One way to locate the polar region is to follow the longer axis of the Cross, as shown here, but there is no obvious marker. The nearest naked-eye star to the pole is a very undistinguished one called Sigma Octantis. On page 121 I will give you details of how to identify it.

To show why an observer's view of the sky depends upon his position on Earth, I suggest that we go on an imaginary journey to the South Pole, in Antarctica — where, incidentally, a major observatory is to be set up within the next few years. Look straight upward to the zenith or overhead point. Reference back to the celestial sphere diagram will show that the south pole of the sky will be directly above us, while the celestial equator will run along the horizon. To make this even clearer, I have redrawn the view on the next page.

As the Earth spins from west to east, the sky seems to turn from east to west around the overhead south pole. The stars will neither rise nor set; each will keep to its own particular distance above the horizon. The stars which lie north of the celestial equator will never come into view at all, so that astronomers at the future South Pole Observatory will be able to examine only half of the sky.

Remember that during its yearly journey along the ecliptic, the Sun spends six months south of the celestial equator and six months in the north. While it is southerly, it will remain above the horizon to our south polar observer, and there will be a 'day' lasting for six months. When the Sun is north of the equator, the south pole will have its six months' 'night'. Obviously exactly opposite conditions apply to the north pole of the Earth; anyone looking at the darkened sky from there will see Polaris directly overhead.

Next, let us travel to the equator (next diagram) and look at the night sky from there. Now, the two poles lie on opposite horizons, while the celestial equator passes overhead; the stars rise due east and set due west if they lie on the equator of the

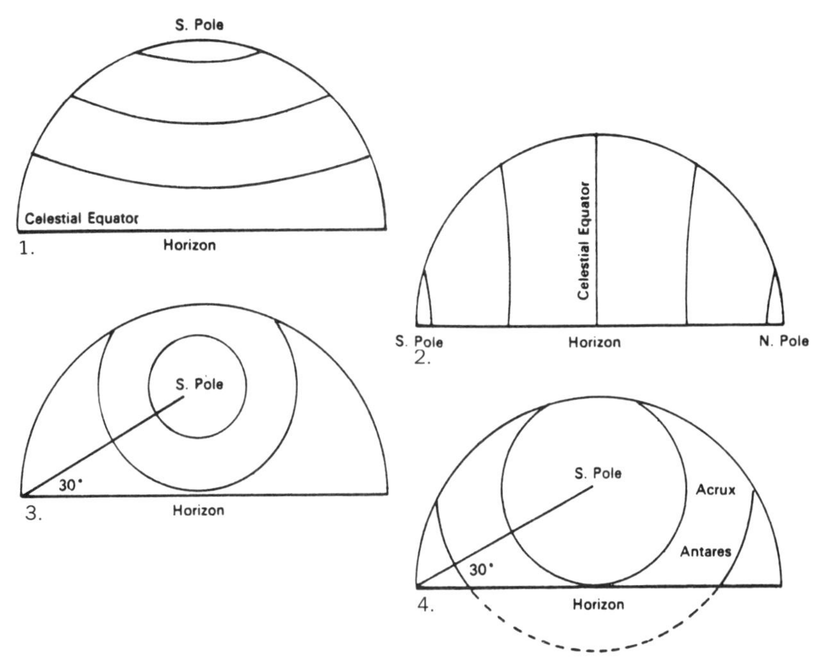

1. View of the sky with the south celestial pole directly overhead.
2. View of the sky from the equator, with the celestial poles on opposite horizons.
3. View of the sky from latitude 30°.
4. Circumpolar stars. From this latitude, Acrux (Alpha Crucis) is just circumpolar, Antares is not.

sky, and both hemispheres can be seen. From Singapore, for example, it always gives me great pleasure to see the two most famous constellations in the sky, the northern Great Bear and the Southern Cross, visible at the same time.

Thirdly, let us come to latitude 40°S. The south celestial pole is high up. Remember, the polar point does not move, because the Earth's axis points straight toward it. The Cross lies not very far from the pole, and so it simply goes round and round without setting. Astronomers say that it is circumpolar, and is always on view whenever the sky is sufficiently dark and cloud-free. Stars further away from the pole do not remain in sight all the time, since for part of the 24-hour circuit they are below the horizon. Thus from Wellington or Sydney, Acrux in the Southern Cross is circumpolar, while Antares, the brilliant red star in the Scorpion, is not (4).

I hope that this chapter has not seemed too dry and technical, but it seemed essential to 'clear the decks' before setting out on a programme of observation. Next, let me say something about the equipment you may feel you need.

Chapter 5

Telescopes And Binoculars

The main astronomical instrument is the telescope, without which we would still know relatively little about the universe. It has to be admitted that a good telescope is not cheap, and very small instruments are emphatically not to be recommended, but it is also true that the cost is non-recurring. Once you have a telescope, it will last you a lifetime provided that you take reasonable care of it.

Telescopes are of two basic types: refractors and reflectors. The refractor collects its light by using an object-glass, which is a special lens (or, rather, a combination of lenses), while the reflector uses a mirror. In either case, the actual magnification of the object you are looking at is done by a second lens called an eyepiece, and every telescope should be equipped with several interchangeable eyepieces to give various powers. In theory, though not always in practice, any eyepiece should fit any telescope.

The diagram shows how the refractor works. The light strikes the object-glass, and the rays are bunched together, meeting at a focus. The image formed is then enlarged by the

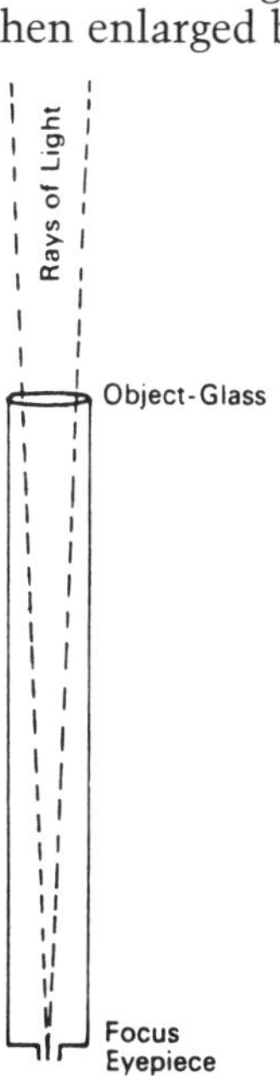

PRINCIPLE OF THE REFRACTOR. Light enters the object-glass and is brought to focus; the image is then magnified by an eyepiece.

eyepiece. The larger the object-glass, the more light can be collected, and the higher the magnification which can be used. Note, by the way, that an astronomical telescope gives an upside-down image. In fact all telescopes will do so, but for an instrument intended for ordinary use, such as bird-watching, an extra lens system is introduced to turn the image the right way up again. However, every time a ray of light passes through a lens it is slightly weakened, and while this does not matter to a bird-watcher it matters very much to an astronomer, who is desperate to collect every scrap of light available. Therefore the correcting lenses are simply left out. It is immaterial whether the Moon or a planet is seen as inverted or erect!

In general, I would advise against paying much money for any refractor with an object-glass below 7.6 cm in diameter. This is equivalent to 3 in imperial measure, and I give it here because of a convenient rule which states that the maximum magnification which can profitably be used is about x50 per inch of aperture. Thus a 3-in or 7.6-cm refractor will give a maximum magnification of 3 x 50, or 150. If you use a more powerful eyepiece, say one which will in theory give a magnification of x250, you will find that the image is so faint that you will be able to see virtually nothing at all.

This leads me on to a warning about an advertiser's trap. You will sometimes find that a telescope is offered for sale because it will say, 'magnify 300 times'. If nothing is said about the diameter of the eyepiece, give the telescope a wide berth. To obtain a power of x300 the object-glass must be at least 6 in or 15 cm across.

The second type of telescope is the reflector, where there is no object-glass. In the usual form — called the Newtonian, because the first instrument of the kind was built by Sir Isaac Newton around 1670 — the light travels down an open tube and

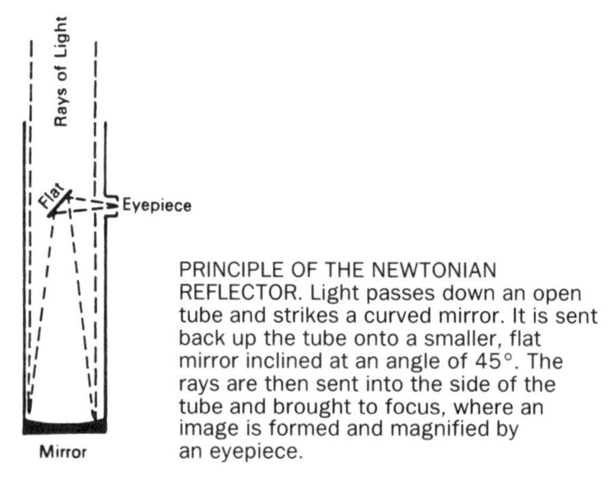

PRINCIPLE OF THE NEWTONIAN REFLECTOR. Light passes down an open tube and strikes a curved mirror. It is sent back up the tube onto a smaller, flat mirror inclined at an angle of 45°. The rays are then sent into the side of the tube and brought to focus, where an image is formed and magnified by an eyepiece.

falls upon a curved mirror at the bottom. The mirror reflects the rays back up the tube onto a smaller, flat mirror placed at an angle of 45°. This flat mirror directs the rays into the side of the tube, where they are brought to focus and the image is magnified as before. With a Newtonian reflector, then, the observer looks into the side of the tube instead of up it.

Aperture for aperture, a refractor is more effective than a reflector, and for a Newtonian I would say that the minimum really useful aperture — that is to say, the diameter of the main mirror — is 6 in or, in round numbers, 15 cm. Of course, smaller telescopes are better than nothing at all, and they can give some pretty views, but they are not likely to satisfy the real enthusiast for long.

On the debit side, a refractor is much more expensive than a Newtonian reflector of equal light-grasp, and it is also less portable. Moreover, it tends to give 'false colour', so that a bright object such as a star will be surrounded by gaudy rings. This is because a beam of light is not so simple as it may seem; it is a blend of all the colours of the rainbow, and when passing through a glass lens the longer wavelengths (red and orange) are bent or refracted less than the shorter wavelengths (blue and violet), so that they are brought to focus in a different place. The false colour nuisance does not apply to a reflector, because a mirror reflects all wavelengths equally. Yet a refractor is much less temperamental than a reflector, and it needs virtually no maintenance, whereas the mirrors of a reflector have to be regularly re-coated with some reflective layer (usually aluminium) and are always liable to go out of adjustment.

There are many other optical systems, some of which are excellent and which make use of a combination of mirrors and lenses. These have most of the advantages of both refractors and reflectors, but tend to be very costly.

A sturdy mounting for the telescope is essential, as otherwise the object under observation will jerk around in an infuriating manner. Some small telescopes are sold upon mountings which are about as rigid as blancmanges. Two movements have to be considered: up and down (altitude) and east to west (azimuth). The simple altazimuth mounting means that the telescope can be moved in any direction. This sounds encouraging, but there is a major problem, because a celestial object is moving across the sky all the time — and if you are enlarging the image by, say 200 times, you are also magnifying the movement by 200 times so that the object will drift quickly across the field of view. The remedy is to have an equatorial mounting, as shown in the next diagram. The telescope is mounted upon an axis which points to the celestial pole, so that when it is moved in azimuth (east to west) the up or down

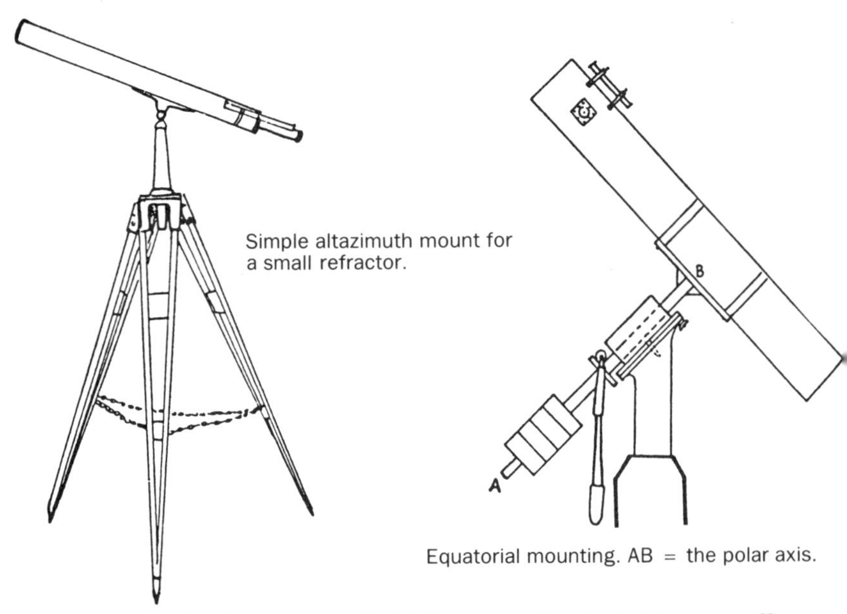

Simple altazimuth mount for a small refractor.

Equatorial mounting. AB = the polar axis.

movement looks after itself. Adding a mechanical drive, usually electrical, means that once the object is in the field, it will stay there — always provided that the mounting has been accurately set up. If you want to take photographs through your telescope, you will have to use time exposures, and an equatorial mounting is almost essential.

If you live in or near a town with glaring artificial lights, you may decide that your telescope must be portable. In this case a 'folded' reflector is ideal — if it is within the aspiring owner's financial limit! If you have a dark site, an observatory is a great help, because the telescope can be permanently set up and there is no need to make adjustments before each session. An observatory may be a simple run-off shed, or a proper dome with a revolving upper section. Leaving a large telescope out in the open is not a good idea, and it is usually wise to dismantle the optics and take them safely indoors when you have finished observing, but there are hazards here too, because it is really only a matter of time before something gets dropped.

Lens-making is beyond anyone apart from the well-equipped expert, but making the mirror for a Newtonian reflector can be done by anyone who is reasonably 'handy'. It is time-consuming, but it is more laborious than actually difficult, even though the beginner must be prepared for many setbacks and disappointments before making a really good mirror. There are books giving all the details, and you will almost certainly find a member of a local or national society ready to give you on-the-spot advice. Of course, another permutation is to buy the optics for a reflector and then construct the mounting, as many amateurs do. Making an observatory is also practicable, be it a run-off shed, a shed with a sliding roof, or a proper dome.

If you cannot afford a telescope, and have no wish to go to the trouble of making one, there is an excellent alternative: equip yourself with binoculars. A pair of binoculars really consists of two small refractors joined together, so that both eyes can be used together instead of only one. Binoculars are defined by the magnification obtained and the aperture of each object-glass, in millimetres; thus a pair of 7 x 60 binoculars yields a magnification of 7, with each object-glass 60 mm across. Binoculars have most of the advantages of a very small telescope except sheer lack of magnification. They will not show the rings of Saturn or the moons of Mars, but they will produce superb views of the craters of the Moon, the phases of Venus, star-fields, coloured stars, clusters and nebulae. If you mean to buy only one pair, then do not select a magnification of more than about x12; if you do, the field of view will be small and the binoculars will be heavy, so that some sort of stand or neck mounting is needed. I would recommend a power of between x7 and x10, with aperture 30 to 60 mm.

I realize that these notes are very sketchy, but all I can do here is to give some general advice, so let me sum up the situation as concisely as I can.

1. Always avoid a very small telescope, below 7.6-cm aperture for a refractor or 15 cm for a Newtonian reflector. Frankly, they are a waste of money from the astronomical point of view. If you do not want to spend a much larger sum, then buy a pair of binoculars.

2. Beware of magazine advertisements. Never buy a telescope in which the magnification is given without any note of the aperture. If you find out that it makes exaggerated claims for more than x50 (7.6 cm) per inch of aperture, avoid it completely. Also, look out for any 'aperture stop' inside the tube which in effect reduces the aperture still further. It is unwise to buy a second-hand or advertised telescope without either seeing it or, preferably, having it checked by an expert.

3. If you want a really powerful telescope at the lowest possible price, consider making yourself a Newtonian, either by buying the optics and mounting them or else making the entire instrument — which is not something to be undertaken lightly, though it is within the scope of anyone who is less clumsy with their hands than I am. Make sure that you have a sturdy mounting for the telescope, whether it be altazimuth or equatorial. If you are making the mounting yourself, I can give you what I believe is a piece of sound advice. Work out the maximum weight which the mounting must have if it is to be

really rigid. When you have done so, multiply by three!

4. If you have no financial constraints, and want a good portable telescope, consider a 'folded' instrument such as a Schmidt-Cassegrain, a Celestron or a Meade.

5. If you have a refractor or a Newtonian of fair size, think about an observatory; it can be a run-off shed, a run-off roof arrangement, or a dome. Make sure that you build it in the best possible position, where the view is not obscured by inconvenient trees or affected by artificial lights.

Once you have a telescope, or a pair of binoculars, you can start to do some really systematic observation. But even if you have to depend entirely upon the naked eye, there is still much to interest you, and it does not take long to learn the language of the sky.

Chapter 6

Our Star: The Sun

In starting our survey of the various bodies in the sky, where better to begin than with the Sun? It is our own particular star, and we owe everything to it; without the Sun, there would be no Earth — and no you or me.

It is extremely large, with a diameter of 1,392,000 km; you could pack more than a million globes the volume of the Earth inside it and still have plenty of room to spare. It is also extremely hot, and this is one reason why it can be dangerous from the viewpoint of an unwary observer — a point to which I will return later. Even the surface temperature is not far short of 6000°C, and near the core, where the energy is being produced, the temperature rises to the incredible value of 14,000,000°C, perhaps rather more.

The bright surface which we can see is called the photosphere. It is made up of gas, and on it we can see the famous dark patches known as sunspots. But before going any further it may be as well to say something about the way in which the Sun shines, and here we must bring in the second of the astronomer's main research tools, the spectroscope.

Just as a telescope collects light, so a spectroscope splits it up. As we have noted, a beam of light is not so simple as might be thought, and is a blend of all the colours of the rainbow. Pass a beam of sunlight through a glass prism, as shown in the diagram, and you will obtain a solar spectrum, made up of a rainbow band — from red at one end through orange, yellow, green and blue to violet — crossed by dark lines. It is these dark lines which enable us to find out what materials are present in the Sun. They are still often called Fraunhofer lines, because the first man to study them in detail, and to map them, was the German optician Josef Fraunhofer, who did so as long ago

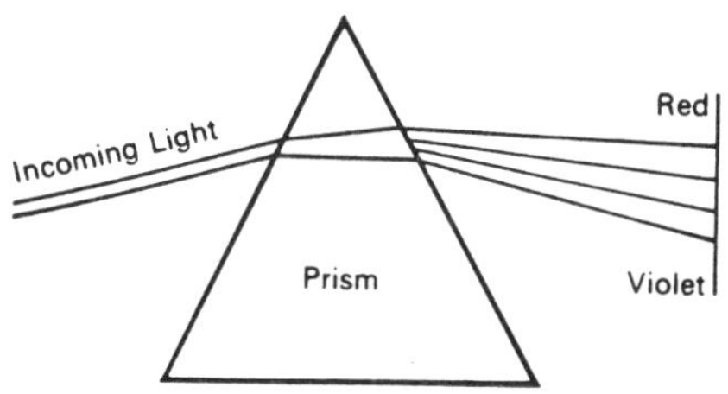

How a prism splits up a ray of light into a rainbow, with red at the long-wave end and violet at the short-wave end.

as 1814. Fraunhofer recorded well over 500 lines, and found that they never changed, either in intensity or in position against the rainbow background. In 1859, long after Fraunhofer's death, another German — Gustav Kirchhoff — interpreted the solar spectrum, and laid the foundations of the modern science of astrophysics.

All matter is made up of atoms, which combine into atom-groups or molecules. Each atom has a central nucleus, around which revolve particles known as electrons. Niels Bohr, a great Danish scientist of our own century, used to picture an atom as a miniature Solar System, with the nucleus representing the Sun and the circling electrons representing the planets. We now know that it is not possible to picture the various parts of an atom as being solid lumps, and the whole situation is much more complicated than we used to think, but Bohr's picture will serve us adequately for the moment. The essential point is that there are only 92 different kinds of atoms known to occur in Nature, and these make up the elements. Typical elements are hydrogen, helium, oxygen, sulphur, iron, tin and gold; all other substances are built up of these 92 basic elements, just as all the words in the English language are built up out of the 26 fundamental letters of the alphabet. For example, water is not an element; its molecule is made up of two atoms of hydrogen together with one atom of oxygen, giving the well-known chemical formula H_2O.

Kirchhoff found that a luminous solid, liquid or high-pressure gas will give a rainbow or continuous spectrum. The Sun's surface consists of high-pressure gas, which explains the rainbow effect. A gas at a lower pressure will not yield a rainbow; instead there will be an emission spectrum, made up of isolated bright lines, each of which is due to one particular element or group of elements. Sodium, one of the two elements making up common salt (the other is chlorine) will yield a whole set of lines, including two which are yellow in colour and lie side by side. If these lines appear, they must be due to sodium, because they cannot be duplicated by anything else. In fact, each element has its own particular trade-mark.

Now let us go back to the spectrum of the Sun. Here we have a rainbow, due to the bright surface. Above the surface of the Sun there is an 'atmosphere' of gases at lower pressure, and these would be expected to produce bright lines. This, as we have noted, is not what we see; there are lines indeed, but they are dark rather than bright.

Kirchhoff found the solution. If a normally bright line is seen against a brilliant rainbow background, it will be 'reversed', and will appear dark. Its position and intensity will be unaltered, and so its origin can be tracked down. In the spectrum of the

Sun, the twin D lines in the yellow part of the rainbow correspond exactly to the two bright yellow lines produced by sodium. Therefore, there must be sodium in the Sun.

Simply by passing sunlight through a spectroscope, we can find out which elements are present; in other words, we can find out 'what the Sun is made of'. The same applies to the stars, which are themselves suns. We can do more than this; we can find out the luminosities, the temperatures and even the sizes of the stars, and we can also find out their distances from us. Without spectroscopes, the modern astronomer would be hopelessly handicapped.

The most plentiful element in the universe is also the lightest: hydrogen, whose atoms more than outnumber those of all the other elements combined. The Sun is made up of more than 70 per cent of hydrogen, and this acts as its 'fuel'. Deep down near the core, where the temperatures and pressures are so colossal, the nuclei of hydrogen atoms are running together to make up nuclei of the second lightest element, helium.* It takes four 'bits' of hydrogen to make one 'bit' of helium, and each time this happens a little energy is set free; also, a little mass is lost, because the four original hydrogen nuclei combined 'weigh' slightly more than the resulting nucleus of helium. It is this energy which keeps the Sun shining, and the loss in mass is four million tons per second, so that the Sun 'weighs' much less now than it did when you started reading this page. Please do not be alarmed, because there is plenty of mass left; the Sun is well over 4,500 million years old, and it will not change much for several thousands of millions of years in the future. This is lucky for us, because even a relatively slight change in the solar output would make the Earth uninhabitable.

Eventually, of course, the supply of available hydrogen 'fuel' will be exhausted, and the Sun will have to change its structure. The outer layers will expand and cool, while the interior will shrink and heat up. Different reactions will begin, and for a while the Sun will become a red giant star, as Betelgeux in Orion is now. Later still the outer layers will be blown away, and the Sun will be reduced to a small, very dense body in which the broken parts of atoms are packed closely together with almost no waste of space. Finally, the Sun will lose all its light and heat, and will turn into a cold, dead globe.

I will return to this theme later on, when we come to discuss stellar evolution in general. Remember, the Sun is a perfectly normal star. But for the moment, let us concentrate upon the Sun as it is today.

* Helium comes from 'helios', the Greek word for 'Sun', because helium lines were found in the solar spectrum long before the element was identified on Earth.

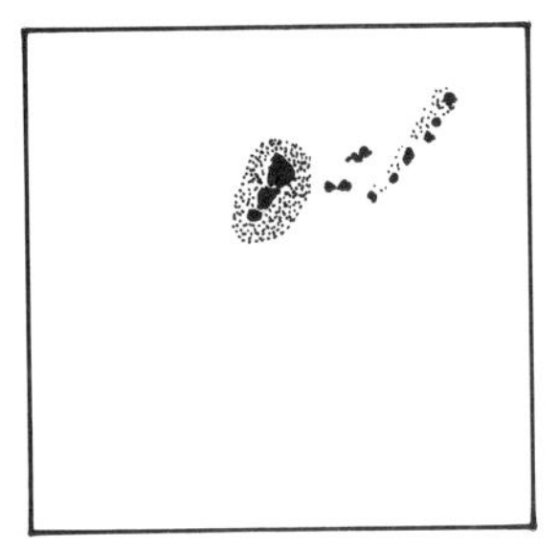

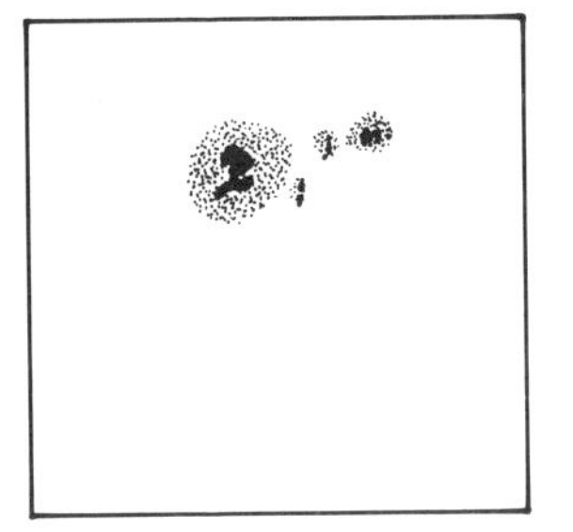

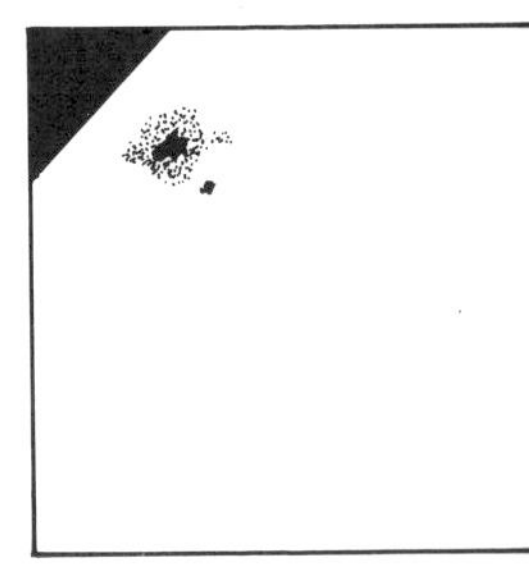

21st, 10h. 24th, 9h. 27th, 9h.

Sunspots, May 1992. Patrick Moore, 12.7-cm refractor x 100. On the 27th, the group was nearing the limb.

On the photosphere we can often see sunspots, which are not really black. They are around 2,000° cooler than the surrounding regions, and look dark by contrast, though if they could be seen shining on their own their surface brightness would be greater than that of an arc-lamp. They are associated with strong magnetic fields, and do indeed appear in places where the Sun's lines of magnetic force break through to the surface, cooling it. They are not always present, and they do not last for very long. A really major group may persist for months, but a small spot may have a lifetime of only a few hours.

Single spots are common enough, but generally we see groups, often with two main members: a leader and a follower. A large spot has a dark central area or umbra, surrounded by a lighter penumbra. Sometimes the shapes are regular, sometimes remarkably complex, with many umbrae contained in one penumbral mass. The Sun is spinning round, and this means that a spot-group can be tracked as it is carried across the Earth-turned face from one limb to the other. The mean rotation period of the Sun is 25 days, so that it takes rather less than a fortnight for a spot to be carried right across the disk. It will then be out of view for a further fortnight, after which it will reappear on the opposite limb if, of course, it still exists. Spots near the limb appear foreshortened, and with a regular spot there is often what is termed the 'Wilson effect', with the penumbra to the limbward side being broadened, indicating that the spot is a hollow rather than an elevation, though not all spots show the effect.

This is where I must give a very serious warning. On no account look straight at the Sun through any telescope or pair of binoculars, even with a dark cap fitted over the eyepiece. If you do, you will focus all the light and — worse — the heat on to your eye, and the result is almost certain to be permanent blindness. This is not mere alarmism; unfortunately it has happened many times in the past, and even a second or two's exposure will be disastrous. No dark cap can give proper protection, because it cannot block out all the dangerous radiations, and is always liable to splinter without the slightest warning.

It is regrettable that some small telescopes (usually refractors) are sold together with dark 'sun-caps'. If you have one of these caps, my advice is to throw it away. In fact, there is only one golden rule about looking straight at the Sun with any telescope: *don't.*

Luckily there is an alternative. Point the telescope at the Sun, without putting your eye anywhere near, and then hold or fix a white card or screen behind the eyepiece. The Sun's image will then be projected on to the screen, showing any spots which happen to be on view as well as the granular structure of the photosphere, due to rising and falling columns of gas. Plotting the spots from day to day, and watching them as they shift and change, is quite fascinating.

The Sun is to some extent a variable star. Every 11 years or so it is very active, with many spots and spot-groups; activity then dies down, and at solar minimum there may be many days with no spots at all, after which activity gradually builds up again toward the next maximum. The cycle is not perfectly regular, and 11 years is only an average, but it is at least a reasonable guide. The last maximum occurred in 1990, so that the next is due around 2001, and the mid-1990s may be expected to be 'quiet'.

It would be idle to pretend that we have a full understanding of the solar cycle, or indeed of the way in which sunspots are produced, but at least we can keep watch on them, together with the bright areas or 'faculae' higher up above the photosphere. These faculae are generally associated with spot-groups. Active groups may also produce violent, short-lived outbreaks known as flares, generally detectable only with special equipment. Flares emit charged particles, which cross the 150-million km gap between the Sun and the Earth and plunge into the upper air, causing it to glow and produce the lovely aurorae or polar lights; aurora australis in the southern hemisphere, aurora borealis in the north.

Actually, the production of auroral displays is not so straight forward as we used to believe, but certainly their origin lies in the Sun. Because the particles are electrified, they tend to spiral downward toward the magnetic poles, and aurorae are best seen in high latitudes. They are not uncommon from Invercargill, for example, but they are less frequent in Auckland, and from Northern Australia they are extremely rare. They are sometimes seen from South Africa — better from the Cape than from Johannesburg, obviously — and close to the equator they are very uncommon indeed, though it is on record that on one occasion an aurora was seen from Singapore.

Aurorae may take many forms. There are glows; arcs, with or without rays; bands, more diffuse and irregular; draperies, or curtains made up of very long rays; and streamers. The colours may be vivid, and the displays shift and change with bewildering speed. If you want to see them in full glory, I recommend a trip to Antarctica, but when the Sun is active the chances of seeing good displays from the South Island of New Zealand at least are fairly good. Certainly the Maori were familiar with them, as we have already noted.

Above the Sun's bright surface lies a layer known as the chromosphere. Ordinary telescopes will not show it, but there are times when it can be seen with the naked eye — when we are treated to the spectacle of a total solar eclipse.

The Earth moves round the Sun; the Moon moves round the Earth. Therefore, there must be times when the three bodies line up, with the Moon in the mid position. When this happens, the Moon blots out the Sun, and by a lucky coincidence (so far as we know, it is nothing more), the two bodies appear almost exactly the same size in the sky; the Sun's diameter is 400 times that of the Moon, but the Sun is also 400 times further away.

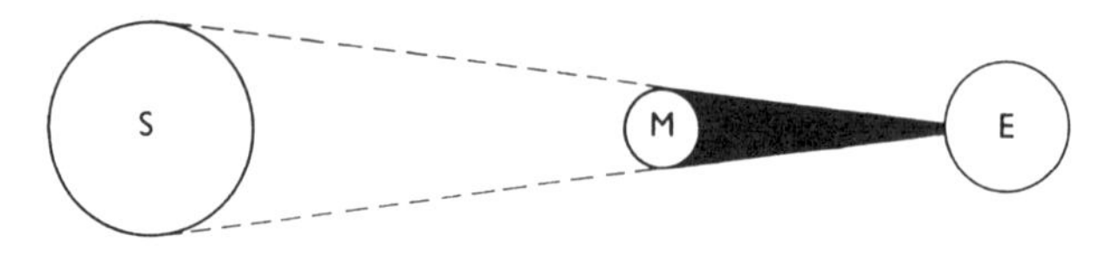

Theory of a solar eclipse. S = Sun, E = Earth, M = Moon.

Solar eclipses are of three types: partial, annular and total. During a partial eclipse, only a portion of the Sun is hidden, and although this is interesting it is not particularly spectacular. At an annular eclipse, the exact lining-up occurs when the Moon is at its greatest distance from the Earth, and appears slightly smaller than the Sun, so that a ring of sunlight is left showing around the dark lunar disk (Latin *annulus,* a ring). When the eclipse is total, the Sun's outer atmosphere flashes into view, and the sight is magnificent. There is the chromosphere, in which can be seen red masses of glowing hydrogen gas known as prominences; beyond comes the corona, a 'pearly mist' stretching outward in all directions. Sometimes the corona is fairly symmetrical; sometimes it sends out long rays.

Because the Moon's shadow is only just long enough to touch the Earth, eclipses as seen from any particular location are rare; one has to be in just the right place at just the right time. Moreover, totality can never last for as long as eight minutes, and is usually much shorter than this. Eclipses between 1994 and 2000 are follows:

1994 May 10	Annular (Pacific, Mexico, USA, Canada)
1994 Nov 3	Total (Peru, Brazil, S.Atlantic). Duration 4m 23s
1995 Apr 29	Annular (Peru, S.Pacific, S.Atlantic)
1995 Oct 24	Total (Iran, India, E.Indies, Pacific). Duration 2m 10s
1996 Apr 17	Partial (Antarctic) 88% obscured
1996 Oct 12	Partial (Arctic) 76% obscured
1997 Mar 9	Total (Siberia, Arctic). Duration 2m 50s
1997 Sept 2	Partial (Antarctic) 90% obscured
1998 Feb 26	Total (Pacific, Atlantic). Duration 4m 8s
1998 Aug 22	Annular (Indian Ocean, E.Indies, Pacific)
1999 Feb 16	Annular (Australia, Pacific)
1999 Aug 11	Total (England, France, Turkey, India). Duration 2m 23s
2000 Feb 5	Partial (Antarctic) 56% obscured
2000 Jul 31	Partial (Arctic) 60% obscured
2000 Dec 25	Partial (Arctic) 72% obscured

Using spectroscopic equipment, it is now possible to study the prominences at any time, but to see the corona well it is necessary either to wait for a total eclipse or else take a trip in a space-craft. The corona is immensely rarefied — millions of times less dense than the air we breathe — and it is also at a high temperature of over a million degrees, but this does not mean that it is 'hot' in the usual sense of the term. Scientifically, temperature depends upon the speeds at which the various atoms and molecules are moving around; the greater the speeds, the higher the temperature. In the corona the speeds are very high, and so is the temperature, but there is very little mass. The best analogy I can give is to compare a firework sparkler with a glowing poker. Each spark is white-hot, but it is of such slight mass that the firework can safely be hand-held — while I, for one, would hate to grasp a red-hot poker, even though the actual temperature is much less!

It is only to be expected that the Sun sends out radiations all through the electromagnetic spectrum, from the very short X-rays through to radio waves. It also emits streams of low-energy particles all the time, making up what is termed the solar wind. It is also a source of strange particles called neutrinos, which are difficult to detect because they have no electrical charge and virtually no mass (perhaps none at all). There seem to be fewer neutrinos than theory predicts, and this is another problem which we have yet to solve.

Solar observation is riveting, and is something which can be done by day even with a telescope set up in the middle of a town. Project the image by all means, and enjoy following the sunspots and faculae. Watch out for the Southern Lights, and if you have the chance to see a total eclipse on no account pass the opportunity by. But let me repeat my warning once more, because it is so vitally important. Even when the Sun is low over the horizon, and looks deceptively mild and harmless, never look straight at it with any optical equipment. A cat may look at a king, but nobody, astronomer or otherwise, should take liberties with the Sun.

Chapter 7
Exploring The Solar System

Less than a century ago, a wealthy French widow provided a substantial sum of money to found what became known as the Guzman Prize. It was to be awarded to the first person who established contact with beings on another planet — but Mars was specifically excluded, because communicating with the Martians would be too easy!

Alas, the Guzman Prize has yet to be claimed (whether those French francs are still stored in a Paris bank I do not know), and we have found that there are no intelligent beings on Mars, or for that matter on any planet in the Solar System with the possible exception of the Earth. By now we have sent space-craft to all the planets except Pluto, and although they are fascinating in their different ways it cannot be said that any of them are particularly welcoming.

First and foremost there is the Moon, less than 400,000 km away and keeping company with us as we travel round the Sun. Accounts of lunar voyages go back a long way, and date back to the second century AD with a story written by a Greek satirist, Lucian of Samosata, whose sailor heroes were hurled onto the Moon when their ship was caught in a violent waterspout. Probably the first serious story along the same lines was Jules Verne's *From the Earth to the Moon*, which came out in 1865 and is still well worth reading. Here, the travellers were sent moonward in a projectile fired from the barrel of a powerful gun.

Verne was not a professional scientist, but he believed in keeping to the facts as far as possible, and at least he dispatched his travellers at the correct speed — 11.3 km per second. This is the Earth's escape velocity. Start off at this speed, and you will not return, because the Earth's gravity will not be strong enough to pull you back. If your velocity is less than this, you will be unable to break free.

Unfortunately, so much friction would be set up by a projectile departing at 11.3 km per second that the result would be a fireball, quite apart from the fact that the shock of such a sudden departure would in any case turn the luckless astronauts into jelly. The only way to go into space is to use the power of the rocket. This was first realized in the 1890s by a shy, deaf Russian school-teacher who rejoiced in the name of

Konstantin Eduardovich Tsiolkovskii. He never fired a rocket in his life, but in many ways his early papers were decades ahead of their time, even though they caused absolutely no general interest when they first appeared.

Consider a rocket of the type used in firework displays. It is made up of a cardboard tube, filled with gunpowder and a stick to give stability in flight. When you 'light the blue touch-paper and retire immediately' the gunpowder starts to burn; it gives off hot gas, and this gas rushes out of the tube through the exhaust. As it does so, it 'kicks' the tube in the opposite direction, and here we have a demonstration of Isaac Newton's principle of reaction: 'Every action has an equal and opposite reaction'. As long as the gas keeps on streaming out, the rocket will continue to fly. Tsiolkovskii also realized that it would be possible to mount one rocket on top of another, so that the upper vehicle could be given what may be termed a running start.

All this was feasible enough, and Tsiolkovskii also pointed out that solid fuels such as gunpowder could be replaced by liquids, making a proper rocket motor. Two different liquids such as petrol and liquid oxygen are pumped into a combustion chamber, where they mix, ignite and produce the gas needed for reaction propulsion. The first liquid-propellant rocket was fired in 1926 by an American, Robert Hutchings Goddard. During the war the Germans developed their V2 weapons for bombarding England, and these V2s were the direct ancestors of the space-craft which have since taken men to the Moon and sent unmanned vehicles out as far as Neptune.

A rocket does not depend upon having air around it, as an aircraft does (even a jet). In fact, air is actually a nuisance, because it sets up friction and has to be pushed out of the way. Moreover, the rocket can start off slowly and work up to full escape velocity only when it is out of the dense atmosphere and is no longer in danger of being burned away by friction.

The Space Age began on 4 October 1957, when the Russians sent up their first artificial satellite or man-made moon, Sputnik 1. Four years later Yuri Gagarin made the first foray into space, completing a full circuit of the Earth before landing safely. And only 12 years separated Gagarin's brief 'hop' and the first manned landing on the Moon.

The first vehicles to reach the Moon were the Russian 'Luniks' or Lunas of 1959. The first of these passed by the Moon, and sent back interesting information, such as confirmation that the Moon has no detectable overall magnetic field. (Go there, and your magnetic compass will not work.) Luna 2 crash-landed, and Luna 3 went right round the Moon and sent back the first pictures of the far side, which can never be seen from Earth because it is always turned away from us (I will have more to say about this in Chapter 8). Subsequently both the Americans and the Russians sent up numbers of unmanned probes, some of which came down gently onto the lunar surface and proved that it was firm enough to bear the weight of a space-craft. Remember, too, that the Moon is much less massive than the Earth. Put the Earth in one pan of a gigantic pair of scales, and you will need 81 Moons to balance it — and it does not pull so hard. On the Moon you will have only one-sixth of your Earth weight, and the escape velocity is a mere 2.4 km per second.

The first manned mission to the Moon was Apollo 11 in July 1969, when Neil Armstrong made his never-to-be-forgotten 'one small step' onto the barren surface of the waterless Sea of Tranquillity.* Five more successful missions followed, the last of which was Apollo 17 in 1972. Since then there have been three unmanned Russian missions, plus a preliminary Japanese foray in 1993.

In 1994 a new American probe, Clementine, spent two months in orbiting the Moon, mapping the surface (notably the polar regions).

Travel to the planets is much more of a problem, because of the increased distance and because the planets, unlike the Moon, do not stay conveniently close to us. Fuel is always the main difficulty, and the solution is to use the Sun's gravity, 'coasting' unpowered for most of the journey.

Again the earliest attempts were Russian, but the first successful interplanetary probe was America's Mariner 2, launched on 27 August 1962. (In case you are wondering what happened to Mariner 1, I have to tell you that it fell in the sea a few minutes after take-off.) Mariner 2 was aimed at Venus, which is the closest of the planets even though it is always at least a hundred times as far away as the Moon. The procedure was to send up the spacecraft in a compound or step-rocket, à la Tsiolkovskii, and then 'slow it down' by using its rocket motors. This would make Mariner start to swing inward toward the Sun along what is known as a transfer orbit. Everything went well; Mariner reached the orbit of Venus on 14 December, after a journey time of nearly four months, and by-passed the planet at a distance of 34,833 km. It carried no cameras, but it was able to send back an amazing amount of information, proving that Venus is emphatically not the sort of world upon which one can spend a quiet week-end.

More than two dozen vehicles have since been sent to Venus; some have made controlled landings there, while others have been put into closed paths round the planet and have

* In a way I rather feel that I span the ages. The first man to fly in a heavier-than-air machine was Orville Wright; the first man in space was Yuri Gagarin; the first man on the Moon was Neil Armstrong. I know, or knew, them all.

mapped the surface by radar. One space-craft, Mariner 10 of 1973–4, has even been on a mission to the innermost planet, Mercury, by-passing Venus en route.

To reach one of the planets beyond the orbit of the Earth, the procedure is to speed the rocket up relative to the Earth, so that it will swing outward instead of inward. Mars, predictably, was the first target. After several failures, the Americans achieved success in November 1964, when they launched Mariner 4. In July 1965 Mariner flew past Mars at a range of less than 10,000 km, and sent back pictures showing a surface which was barren and cratered — sadly, with no sign of Madame Guzman's Martians! In 1971 Mariner 9 entered a closed path around the planet and sent back detailed images; in 1976 two Viking probes made controlled landings, and were able to 'scoop up' Martian material, analyze it, and transmit back the results. The main aim had been to search for life, but no definite traces of any living organisms were found.

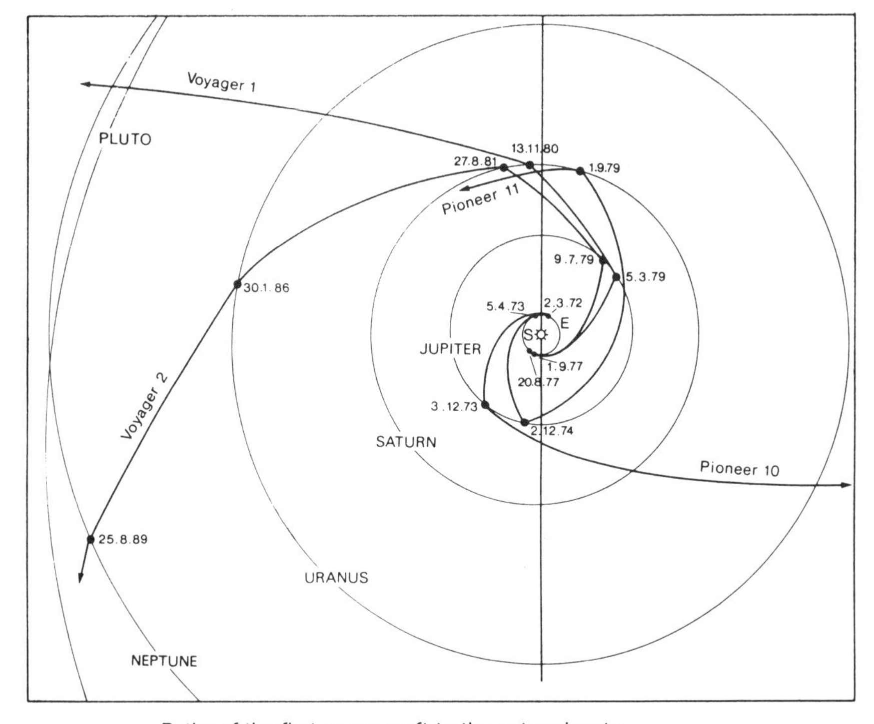

Paths of the first space craft to the outer planets;
Pioneer 10 encountered Jupiter.
Pioneer 11 encountered Jupiter and Saturn.
Voyager 1 encountered Jupiter and Saturn.
Voyager 2 encountered Jupiter, Saturn, Uranus and Neptune.
Unfortunately Pluto has been missed out!

The giant planets are so much further away that a journey there must take years, as against days (for the Moon) or months (for Venus, Mars or Mercury). The first attempt was American. On 2 March 1972 Pioneer 10 was launched, and after successfully negotiating the asteroid belt it by-passed Jupiter on 3 December 1973. It was a complete success, sending back superb pictures as well as invaluable data, but it nearly failed, because the radiation zones around Jupiter turned out to be much more powerful than had been expected, and the instruments on Pioneer were saturated. Pioneer 11 followed a year later; this time the path was modified so that it passed quickly over Jupiter's equatorial zone, where the radiation hazard is at its worst.

The era of improvisation had begun. After by-passing Jupiter, Pioneer 11 was found to have sufficient fuel left to send it back across the Solar System on to an encounter with the second giant planet, Saturn, on 1 September 1979. But by then two much more ambitious space-craft, the Voyagers, were already on their way.

In the late 1970s, by a lucky chance, the four giants — Jupiter, Saturn, Uranus and Neptune — were spread out in a sort of long curve, so that it was possible to send a space-craft to each in turn. This is the procedure known as 'gravity assist', though I always think of it as interplanetary snooker. It is basically a fuel-saving technique, and it has been a great success. Voyager 1 encountered Jupiter first (in 1979) and then Saturn (in 1980), but let us concentrate upon the second Voyager, which had a much more complicated programme and which may qualify as the most successful space-craft ever launched.

Voyager 2 went on its way on 20 August 1977. It was speeded up in the usual manner, and passed through the asteroid belt, reaching the neighbourhood of Jupiter on 9 July 1979. It swung round Jupiter at a minimum distance of 714,000 km, and used the immensely powerful Jovian gravity to propel it on to an encounter with Saturn on 26 August 1981. Saturn's pull was then used to send Voyager on to an encounter with Uranus on 24 January 1986, and finally Uranus itself sent Voyager on to Neptune, which was by-passed on 25 August 1989. Superb pictures and masses of data were sent back from all four giant planets. At the time of the Neptune pass, Voyager 2 had been in space for almost 12 years and had covered over 6000 million km — yet it reached its target within two minutes of the planned time. The distance from Earth was then 4,425 million km, so that even a radio signal took 4 hr 6 m to reach us.

The technical problems were daunting, and it is fair to say that Voyager 2 surpassed all the hopes of its planners. At the present time (1994) we are still in touch with Pioneer 10 and

11 and with both the Voyagers, but they will never come back; they are leaving the Solar System, and before long we are bound to lose contact with them. In millions or even thousands of millions of years hence they may still be travelling between the stars, unseen, unheard and untrackable. They carry records and plaques, so that if any alien civilization picks them up their planet of origin might be identified. though one has to admit that the chances of their being found are not very great.

There have been various other planetary missions since then. Magellan has produced first-class maps of Venus. The ambitious Mars Observer of 1993 failed, but another probe, Galileo, is on its way to Jupiter, which should be reached in 1995. It has already sent back the first close-range asteroid pictures. Neither must we forget the armada of spaceships sent to rendezvous with Halley's Comet in 1986.

What will happen next? Well, a Lunar Base is a real possibility, and there are tentative plans to send the first manned expedition to Mars, though it is too early to give a definite date, and the human body may be the weak link in the chain. During the coming century we will dispatch further unmanned probes, and we will learn much from them.

There are still a few people who are short-sighted enough to question the whole value of space exploration. They forget that the sums of money spent are trifling by national standards (a space-craft to the Moon costs less than a nuclear submarine), and in any case it is no longer possible to separate one branch of science from another. To give just one example: there is a very close link between space research and medical research. A few days ago I visited a hospital, and saw the results of the scanning of an unborn baby carried out with equipment which had originally been developed for use in the space programme.

The fact that rockets can also be used in warfare is regrettable, and it is undeniable that but for Hitler's terror weapon the progress of space research would have been much less rapid than it actually has been. But this is a problem for the politicians, not the scientists, and we can only hope that we will learn our lessons in time. If not, then we will never reach Mars — and neither will we deserve to do so.

Chapter 8

The Moon

The Maori, the Aborigines and the Bushmen saw men, women and animals in the Moon. Even in our own century it was widely believed that there might be low-type vegetation there. Nowadays we know better. The Moon is sterile, and always has been so; the first living beings on the lunar surface were the Apollo astronauts. Yet insignificant though it may be in the Solar System, the Moon is of unique importance to us.

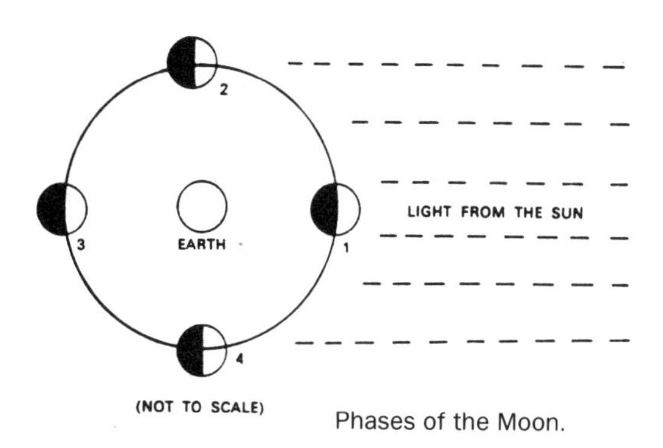

Phases of the Moon.

As everyone knows, the Moon is not always on view. It shows regular phases, or changes of shape, which have been seen ever since the dawn of human history — and even before. The reason for this is easy to understand, and I hope that the diagram will make it perfectly clear.

The Moon moves round the Earth, taking 27.3 days to complete one journey,* and it has no light of its own; it shines only because it is being illuminated by the Sun. In position 1, the dark side of the Moon is turned toward us, and the Moon is new, so that we cannot see it at all. As it moves along in its orbit, a little of the sunlit side begins to be turned toward us, and the Moon shows up, first as a crescent, then as a half (position 2) and then as full (position 3). After that the phase starts to decrease once more, returning to half at position 4 and then shrinking to a crescent. Finally it returns to position 1, and

* To be strictly accurate, the Earth and Moon move together round their common centre of gravity, or barycentre; but this lies deep inside the Earth — because the Earth is 81 times more massive than the Moon — so that the simple statement that 'the Moon goes round the Earth' is good enough for most purposes.

again the Moon is new. Because the Earth-Moon system is moving round the Sun, the mean interval between successive new moons is 29½ days instead of 27.3, but generally speaking we have one new moon and one full moon every month.

When the Moon is at its crescent stage, the 'dark' side may often be seen shining dimly. This is due to light reflected from the Earth on to the Moon, and is therefore known as the earthshine. Incidentally, the Moon is a poor reflector; on average its rocks reflect only seven per cent of the sunlight which falls upon them, so that instead of being dazzling white they are really dull grey. As so often happens in astronomy, appearances are deceptive.

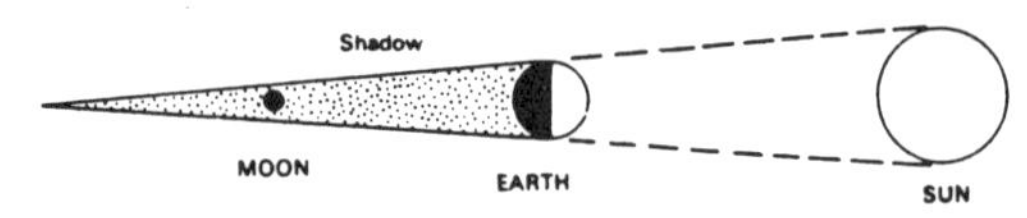

Theory of a lunar eclipse.

If the Earth, Moon and Sun line up, the result is an eclipse — a solar eclipse when the Moon is in the mid position, a lunar eclipse when the Earth lies between the Moon and the Sun. The Moon then passes into the cone of shadow cast by the Earth, and the supply of direct sunlight is cut off, so that the Moon becomes very dim until it passes out of the shadow again. Lunar eclipses may be either total or partial.

Since the Moon depends entirely upon reflected sunlight, it might be thought that it would vanish completely during an eclipse. This does not (usually) happen, because some of the Sun's rays are bent or refracted on to the Moon by way of the shell of atmosphere surrounding the Earth. Not all eclipses are equal, because everything depends upon conditions in the Earth's air. For example the two eclipses of 1993 were exceptionally dark because of the dust and ash in the atmosphere sent out by the eruption of Mount Pinatubo, in the Philippines. It cannot be said that lunar eclipses are astronomically important, but they are beautiful to watch. Between now and the end of the century, total eclipses will occur on 1996 April 4, 1997 September 16, and 2000 January 21 and July 16; partial eclipses on 1994 May 25 and November 18, 1995 October 8, 1997 March 24, 1998 March 13 and August 8, and 1999 January 31 and July 28.

The Moon is a very different world from the Earth. Its low mass and low escape velocity mean that it has lost any atmosphere it may once have had, and there is a complete lack of water. Originally it was thought that the broad dark plains visible even with the naked eye were true seas, and we still call

Main features of the Moon.

them by their romantic names: the Sea of Tranquillity, the Bay of Rainbows, the Ocean of Storms and so on. Astronomers always use the Latin names, and I propose to do so here; there is nothing difficult about them.

The 'seas' are lava plains, and there has never been any water in them. The largest of them, the Oceanus Procellarum (Ocean of Storms) covers an area greater than that of our Mediterranean. Some of them are bounded by ranges of mountains. The most prominent of all the seas is the Mare Imbrium (Sea of Showers), which is more or less regular in outline, and is bordered by lofty mountain ranges such as the Apennines, the Alps and the Carpathians. Some of the peaks in the Apennines tower to over 4000 metres above the surrounding landscape.

The whole of the Moon is dominated by the walled circular enclosures which we call craters, even though in many cases 'walled plains' would be a better description. They range in size from huge enclosures over 200 km across down to tiny pits too small to be seen from Earth even with our most powerful telescopes. No part of the surface is free from them; they cluster thickly in the bright uplands, and are also to be found on the maria and even on the summits of mountains. Not all of them are regular; they break into each other and deform each other, and there are also 'ghost' craters, mainly on the seas, which have been almost obliterated by the mare lava. Here and there we find immense 'bays', where the seaward wall of the old crater has been levelled.

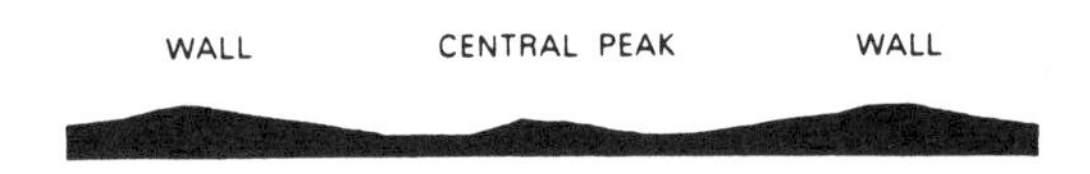

Cross-section of a large lunar crater. It is more like a shallow saucer than a deep mine-shaft.

A typical crater consists of a circular rampart surrounding a sunken floor, on which there may be a central mountain or mountain group. Generally speaking, the depth is not great in relation to the diameter, and the wall-slopes are fairly gentle, so that if drawn in profile a crater resembles a shallow saucer rather than a mine-shaft — though there is one formation, known as Wargentin, which is filled to the brim with lava, so that it has taken the form of a plateau almost 90 km across. Note that the central peak never equals the height of the surrounding walls, and many craters have no central structures at all. Such is the 96-km Plato, near the Mare Imbrium, which is perfectly circular and has a flat, iron-grey floor. As seen from Earth it appears elliptical, because it is well away from the centre of the disk and is considerably foreshortened.

Mountain ranges are generally named after terrestrial ranges, but the craters are named after personalities, usually astronomers, though some non-astronomers have found their way there. (Julius Caesar has a crater of his own, though it was named to honour his connection with calendar reform rather than his military prowess!) The system was introduced by an Italian named Riccioli, who drew a lunar map in the year 1651. Most of his names have been retained, though of course many others have been added since.

The terminator, or boundary between the daylit and night hemispheres, is always rough and uneven; the lunar surface is mountainous, and of course the Sun's rays will shine on a peak while the adjacent valley is still in darkness. A crater is at its most imposing when it is near the terminator, so that part or all of its floor is plunged in shadow. When the Sun rises high over it, the shadows vanish, and the crater may become hard to identify unless it has an exceptionally dark floor (as with

Plato) or exceptionally bright walls and central peak (as with the 37-km Aristarchus, in the Oceanus Procellarum). The aspect of a crater may change dramatically even over a period of a few hours. To my mind, the most beautiful of all lunar scenes is the so-called 'Jewelled Handle', seen when the Sun is rising or setting over the mountainous western border to the Sinus Iridum (Bay of Rainbows) which leads off the Mare Imbrium. The Bay itself is darkened, and the mountains seem to project outward from the terminator.

Near full moon, when there are virtually no shadows, many of the large craters become hard to identify, and the whole scene is dominated by the bright streaks or rays which are centred upon some of the craters, notably Tycho in the southern uplands and Copernicus in the Mare Nubium (Sea of Clouds). The rays are surface deposits, so that they are well seen only when the Sun is high above them; when Tycho is close to the terminator it looks like nothing more than an ordinary, well-formed bright crater, 87 km across, with a central mountain group. The rays do not issue from the exact centre of the crater, but are tangential to the walls; they extend over a large part of the Earth-turned hemisphere. There are also many minor ray-systems.

The Moon spins slowly on its axis. It takes 27.3 days to make one full turn, and this is the same time that it takes to complete one orbit round the Earth. This means that it keeps the same face turned toward us all the time. This can sometimes cause confusion, but it is easy to give a practical demonstration to show what happens. Walk round a chair, turning as you go so as to keep your face turned chairward. When you have completed one circuit, you will have 'turned once on your axis' (you will have faced all four walls of the room) but anyone sitting on the chair will never have seen the back of your neck. Similarly, we on Earth never see the 'back' of the Moon. This also explains why the markings on the lunar disk always keep to the same positions. For example Tycho is always near the southern limb, Plato well to the north, and the well-marked almost circular Mare Crisium (Sea of Crises) to the north-east.

There is, however, one modification. The Moon spins on its axis at a constant speed, but it does not move along in its orbit at constant speed. The orbit is slightly eccentric, and, following the usual traffic laws of the Solar System, the Moon moves quickest when it is closest to us (perigee) and slowest when it is furthest away (apogee). The perigee and apogee distances are respectively 384,400 km and 406,697 km, so that the difference is quite appreciable. Therefore, the amount of axial spin and the position in orbit become periodically 'out of step', and we can see a little way round first one mean limb and then the other. This effect is known as 'libration in longitude'.

Together with various other librations, the result is that we can examine a grand total of 59 per cent of the total surface, though of course never more than 50 per cent at any one time. The remaining 41 per cent is permanently turned away from us — a fact which pre-Space Age astronomers, such as myself, found quite infuriating. Moreover, formations close to the limb are so foreshortened that it is often difficult to tell the difference between a crater and a mountain ridge.

Although the Moon keeps the same face turned toward the Earth (discounting the effects of librations), it does not keep the same face turned toward the Sun, so that day and night conditions are the same everywhere, apart from the fact that from the far side the Earth can never be seen. There is no mystery about the coincidence between the axial rotation period and the orbital period. In its early history the Moon was viscous, and the Earth raised huge tides in it, so that there was a 'bulge' pointing Earthward. This braked the lunar rotation until relative to the Earth, though not relative to the Sun, it had stopped altogether. Most of the main satellites of other planets have similarly captured or 'synchronous' rotations.

Until 1959 we had no direct knowledge of the Moon's far side, but in October of that year the Russians sent their unmanned probe Luna 3 on a round trip, and secured the first pictures. As expected, the far side is just as barren and just as crater-scarred as the area we have always known, and by now we have very detailed maps of the entire surface. On the far side there are some impressive features, such as the circular formation named in honour of Tsiolkovskii, which seems to be a cross between a 'sea' and a 'crater'. The main difference between the Earth-turned and the far hemispheres is that on the far side of the Moon there are no maria comparable with, say, the Mare Imbrium — and as we have already noted, the seas which we can see from Earth form, in the main, a connected system; the Mare Crisium is separate, but none of the major seas extend on to the far side.**

Nobody is confident about the origin of the Moon. It certainly did not break free from the Earth and leave a vast hole now filled by the Pacific Ocean, as was once believed. It may

** The main exception is the Mare Orientale or Eastern Sea. I have fatherly interest in it, because I was the first to report it, long before the Space Age, when I was busily mapping the foreshortened libration areas with the telescope in my modest observatory. I took it to be a minor feature, and not until the rocket era was it found to be a vast, ringed formation, most of which is inaccessible from Earth; all we can see from here is its extreme boundary. I suggested the name for it, because it lay on what was then regarded as the Moon's eastern limb. Then, later on, a resolution of the International Astronomical Union, the controlling body of world astronomy, reversed east and west, so that my Eastern Sea is now on the western limb. I voted against the change, but I was heavily defeated!

have been formed in the same way as the Earth, from a cloud of material surrounding the youthful Sun, but there is a growing feeling that it may have been the result of a collision between the Earth and a wandering planet-sized body around 4500 million years ago. At least we know that the Earth and the Moon are of the same age, because we have been able to date the rocks brought home by the Apollo astronauts and a few unmanned Russian sample-and-return probes.

There has also been a great deal of argument about the origin of the craters. Today most astronomers believe that they were produced by a natural meteoric bombardment, though there are a few people (such as myself) who prefer to regard them as essentially volcanic. Certainly they are ancient by terrestrial standards, and there can have been no major changes on the Moon for well over 1000 million years, probably much longer.

The only activity which we can trace today takes the form of very mild, localized gaseous emissions from below the crust, which have been recorded by persistent lunar observers and are known as TLP or Transient Lunar Phenomena (a term for which I believe I was originally responsible). They are restricted to certain areas, such as that of Aristarchus in the Oceanus Procellarum, which is the most reflective formation on the Moon and can glow so brightly even when lit only by earthshine that unwary observers have been known to mistake it for a volcano in eruption.

Most people know that the initial manned flight to the Moon was made in July 1969, when first Neil Armstrong and then Edwin Aldrin, in the lunar module of Apollo 11, stepped out on to the Mare Tranquillitatis. I doubt if anyone has ever bettered Aldrin's description of the lunar scene as 'magnificent desolation'. Nothing moves, nothing breathes, nothing stirs; the sky is black, as there is no air to smear the sunlight around and make the sky blue; everywhere there are hills, craters and pits. Then too, there is the Earth, shining much more brilliantly than the Moon does to us. If the astronaut shields his eyes from the glare of the rocks, he can see the Sun and the stars at the same time, shining down from the inky blackness of the lunar sky.

Apollo 11 was the first of the series; others followed, and all except No.13 were successful. Regions of all types were explored, and the astronauts of Apollos 15, 16 and 17 even drove around in 'moon cars' which they had brought with them. Scientific stations were set up, and went on transmitting for some years after the last astronauts had left. No doubt they will again be visited in the foreseeable future, and the abandoned moon cars will be driven away to a lunar museum.

We know a great deal more about the Moon now than we did at the start of the Apollo programme. Atmosphere is virtually absent; the lunar surface is covered with a rough layer known as the regolith, below which is firmer rock. The Moon's core is certainly hot, though the temperature is much lower than that at the centre of the Earth. Moonquakes occur, though they are so mild that they will pose no threat to future manned bases. Very little happens on the Moon today; the obligatory American flags set up by the astronauts do not flutter — there is no wind to make them do so — and the footprints left on the surface will remain there until covered up by cosmic dust.

The 1994 lunar probe, Clementine, paid particular attention to the Moon's polar regions, which are very foreshortened as seen from Earth. Some of the polar craters have floors which are permanently in shadow, and are therefore bitterly cold. Clementine suspected the presence of ice there — and if this is confirmed, it will be a very welcome development for future prospective colonists!

It does not take long to learn one's way around the Moon. Even binoculars will show the mountains, seas, valleys and craters, and with a telescope you can also examine the minor features, such as the swellings known as domes, the crack-like clefts or rills, and exceptional structures such as the inappropriately-named Straight Wall, which is simply a fault in the surface. The best views are obtained when the Moon is either a crescent, half or three-quarter, or gibbous, shape. Near full there are almost no shadows, and the familiar features are masked by the Tycho and Copernicus rays.

Lifeless though it may be, the Moon is a world of supreme interest. Even though it has now been visited by men from earth, it has lost none of its magic.

Chapter 9
The Inner Planets

Let me now invite you to join me on an admittedly rather hurried and breathless tour of the Solar System. It will be convenient to start close to the Sun and to move outward, so that our first visit must be to Mercury.

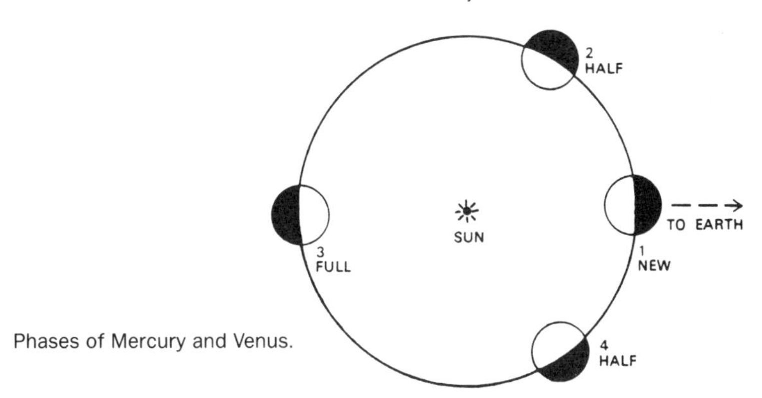

Phases of Mercury and Venus.

Mercury moves round the Sun at a mean distance of 60 million km in a period of 88 days — one Mercurian 'year'. Like the Moon, it shows phases, and for much the same reason. In the diagram it is 'new' at position 1, when its dark side is facing us, and it cannot be seen at all except on the rare occasions when it passes in transit across the face of the Sun. It is a crescent between positions 1 and 2, and half-phase, known technically as dichotomy, when it has reached 2. The phase goes on increasing until position 3, when Mercury is full, but it is then out of view on the far side of the Sun. When it reappears in the evening sky the phase shrinks to half at position 4, then to a crescent and then back to new at position 1.

Transits are interesting, even if not important; Mercury appears as a dark disk against the brilliant solar face. However, Mercury's path is tilted relative to ours, and on most occasions 'new Mercury' passes unseen either above or below the Sun in the sky. The only transits between 1990 and 2000 are those of 6 November 1993 and 15 November 1999.

The fact that Mercury is so close to the Sun makes it an awkward object to observe. It is visible with the naked eye only when best placed, either very low in the west after sunset or very low in the east before dawn, and there must be many people who have never seen it at all. Moreover it is a small world; its diameter is a mere 4878 km, not a great deal larger than that of the Moon — and since it never comes much within 80 million km of us, no telescope will show much on its disk. Before the 1960s we did not even know the length of its 'day', though we now know that it is 58.6 times as long as ours. The Mercurian calendar is decidedly peculiar!

The small mass means that there is practically no atmosphere, and in every way Mercury is a hostile world. One space-probe has encountered it: Mariner 10, which made three active passes in 1974 and 1975, sending back pictures showing a barren, mountainous, crater-scarred landscape, superficially very like that of the Moon. The daytime heat is fearsome. In places it reaches almost 43 °C — but the night cold is equally daunting, because the temperature plunges to below -180 °C. It is very clear that life on Mercury is out of the question, and we have to admit that the planet is not of much interest to the amateur observer.

Planet No.2, Venus, is very different. Here the distance from the Sun is 108 million km. The revolution period is 224.7 Earth days, but Venus' own axial rotation period is 243 Earth days, so that in effect the 'day' is longer than the 'year'. To make matters even stranger, Venus spins from east to west, so that if the Sun could be seen from the planet's surface it would rise in a westerly direction and set in the east 118 Earth days later. In size and mass Venus and the Earth are near-twins.

However, an observer on the surface would never see the Sun at all, because Venus has a dense, cloud-laden atmosphere which would hide the sky completely. From Earth we cannot see through it, and before the Space Age our ignorance of the conditions below the clouds was more or less complete. We did not even know whether Venus was a raging dust-desert, or was largely covered with water. Vague, cloudy shadings appeared at the top of the atmosphere, and we could use spectroscopes to analyze the atmosphere itself, which was found to be quite different from ours; instead of nitrogen and oxygen, the main constituent turned out to be carbon dioxide. Carbon dioxide tends to act in the manner of a greenhouse, and shuts in the Sun's heat, so that at least we could tell that Venus must be very torrid.

(Had there been oceans, the carbon dioxide in the atmosphere would have fouled them, and the result would have been seas of soda-water. This was regarded as quite possible, though I remember commenting that the chances of finding any whisky to mix with it did not seem very promising.)

Mariner 2, the first successful interplanetary probe, by-passed Venus in 1962 and confirmed the very high temperature, so that the attractive marine theory had to be cast onto the scientific scrap-heap. Subsequently the Russians managed to bring several space-craft down gently on to the surface, though none could transmit for very long before being put out of action by the intensely hostile conditions. Quite apart from the heat and the crushing atmosphere, the clouds had been found to be rich in corrosive sulphuric acid. It was greatly to the Russians' credit that several pictures were obtained, showing an orange, rock-strewn surface. Winds were slight, even though the upper clouds spin round much more quickly than the solid body of the planet.

The best maps of the surface have been obtained not by visual observation, but by radar. Probably most people today are familiar with the principle of radar; what is done is to send out a pulse of energy, bounce it off a solid body (or equivalent) and then receive the 'echo', which can provide information about the nature of the object off which the pulse has bounced. (There is some analogy, though not an accurate one, with a tennis-ball which has been thrown against a wall and caught on the rebound; the way in which the ball comes back will tell you at least something about the wall.) Space-craft carrying radar have been put into closed paths round Venus, and they have been very successful, particularly Magellan, which was launched in May 1989, began orbiting Venus in August 1990, and was still transmitting in 1994. Venus proved to be a world of plains, highlands and lowlands, with valleys, craters and volcanoes which are probably active, together with strange features which look superficially rather like spider's webs and have been called 'arachnoids', circular volcanic structures surrounded by complex details.

One of the smaller highlands, Beta Regio, contains two massive peaks which have been named Rhea Mons and Theia Mons. Theia is of the same type as our Hawaiian volcanoes, but is much more massive. We cannot prove that it is erupting today, but it does seem very likely. Certainly there is evidence of vulcanism all over the planet, and there are extensive lava-flows.

If Venus and the Earth are near-twins in size and mass, we might expect them to be in the same state, but nothing could be further from the truth. The difference seems to be due to the fact that Venus is over 30 million km closer to the Sun than we are. It is thought that in the early days of the Solar System the Sun was less luminous than it is now, and presumably Venus and the Earth started to evolve along the same lines, developing the same kinds of atmospheres and the same kinds of seas. Then, gradually, the Sun became hotter. Earth was just out of the danger-zone. Venus was not, so that the oceans boiled away, the carbonates were driven out of the rocks, and in a relatively short time Venus changed from being a potentially life-bearing world into the furnace-like inferno of today.

As seen with the naked eye, Venus is glorious; at times it may cast strong shadows. Telescopically, however, little can be seen apart from the characteristic phase, together with the vague, cloudy shadings which shift and change. It was natural for early peoples to name Venus in honour of the Goddess of Beauty, but we now know that conditions there approximate much more closely to the conventional idea of hell.

Passing by the Earth and its Moon, we come next to Mars, the planet whose strong red colour led to its being named after the mythological God of War. In size and mass Mars is intermediate between the Earth and the Moon; the diameter is 6794 km, and the mass one-tenth of ours, giving an escape velocity of 5 km per second. We might therefore expect a thin but appreciable atmosphere, and this is just what we find. The ground pressure is below 10 millibars everywhere, and the main constituent is carbon dioxide, so that beings such as you and me certainly could not breathe there.

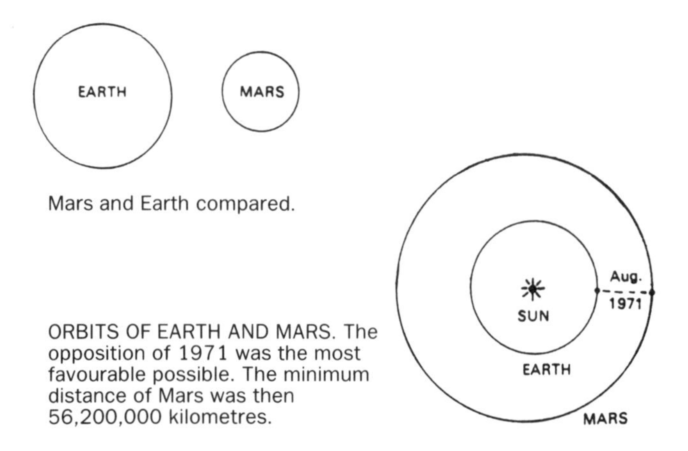

Mars and Earth compared.

ORBITS OF EARTH AND MARS. The opposition of 1971 was the most favourable possible. The minimum distance of Mars was then 56,200,000 kilometres.

A modest telescope will show surface markings when Mars is well placed; there are dark patches, ochre regions which are still called 'deserts', and white ice-caps covering the poles. At least we are seeing true surface features, not mere clouds as with Venus.

Mars is best seen when it is opposite to the Sun in the sky, and is therefore due north at midnight; this is termed 'opposition'. Oppositions of Mars occur every alternate year; thus there were oppositions in 1993, 1995 and 1997, but not in 1994 or 1996. The next diagram shows the cause of this

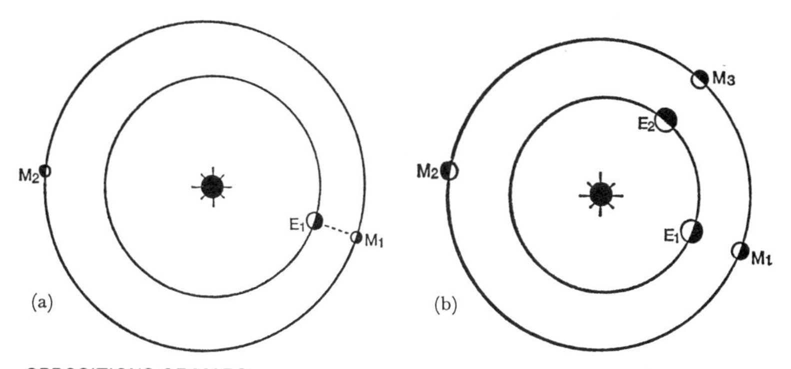

OPPOSITIONS OF MARS.
(a) Opposition; Earth at E1, Mars at M1. One year later the Earth is back at E1, but Mars has only reached M2.
(b) The next opposition; Earth at E2, Mars at M3.

behaviour. Begin with the Earth at E1 and Mars at M1; clearly this is the time of an opposition. One year later the Earth has completed a full circuit of the Sun, and has arrived back at E1, but Mars has not had time to do so. It is moving more slowly in a larger orbit — its mean distance from the Sun is 228 million km and its revolution period is 687 Earth days, so that it has only reached M2. The Earth has to catch it up, so to speak, and the next opposition falls with the Earth at E2 and Mars at M3. The average interval between one opposition and the next (the synodic period) is 780 days.

Though the Martian year is so much longer than ours, it has a day which is of the same order; 24 hours 37 minutes, so that there are 669 Martian days or 'sols' in every Martian year. The tilt of the axis is about the same as ours, so that the seasons are of the same type, though much longer.

There is another complication, too. Mars has an orbit which is much less circular than ours, and the most favourable oppositions occur when the planet is at its closest to the Sun, as in 1988 and 2003. At its closest, Mars may approach us to a distance of 58,400,000 km, which is roughly 150 times as far away as the Moon. Note, incidentally, that southern summer on Mars falls when the planet is near perihelion, so that the southern summers are shorter and hotter than those of the northern hemisphere while the southern winters are longer and colder.

The dark areas on the surface are permanent, and near opposition a small telescope will show some of them, such as the rather V-shaped feature once known as the Hourglass Sea and now known as the Syrtis Major. The ochre tracts cover much of the disk; the polar caps wax and wane with the Martian seasons, so that in winter they are large and in summer become almost or quite invisible.

Originally it was thought that the dark areas were true seas, but gradually it became clear that the Martian atmosphere is too thin for oceans to exist, and it was more generally felt that the dark areas must be old sea-beds filled with vegetation. The ochre tracts were classed as deserts, though they are cold rather than hot and are covered with coloured minerals rather than sand; the caps were assumed to be ordinary water ice, though possibly no more than a very thin layer of frost. Chilly though it certainly was, Mars was clearly much less hostile than Mercury or Venus, and before 1965 most astronomers were confident that there was low-type life there, even if the 'Martians' belonged to the pages of science fiction novels. (Less than a century ago, some observers were still drawing fine, artificial-looking lines which were called canals, and which were attributed to intelligent beings who had built a planet-wide irrigation system. Alas, we have now found that there are no canals; they were due simply to tricks of the eye.)

The first successful Mars probe, Mariner 4, made its pass in 1965, and others have followed, with the result that most of our cherished ideas about the planet have had to be drastically revised. The red tracts are indeed deserts covered with reddish minerals, but the dark regions are not depressions and they are not covered with vegetation. Some of them, such as the Syrtis Major, are lofty plateaux, and they are merely areas where the reddish material has been blown away by winds in the thin atmosphere, exposing the darker surface beneath. The polar caps are made up of a combination of water ice and carbon dioxide ice, and they are many metres thick. The atmosphere itself turned out to be much less dense than had been expected; as we have noted, it is made up chiefly of carbon dioxide, and the pressure at ground level is much less than that in the Earth's air at three times the height of Mount Everest. This indicates that any advanced life-forms are out of the question, which was frankly disappointing.

Two space-craft, Vikings 1 and 2, made controlled landings on Mars in the winter of 1975, and sent back pictures showing a red, rocky surface under a pink sky. Material was scooped up, drawn into the space-craft and analyzed. The results showed no definite sign of life of any sort, and it is now thought likely that there is no living thing on Mars, at least at the present time.

One major surprise was the presence of craters, mountains, deep canyons and lofty volcanoes. One volcano, now named Olympus Mons (Mount Olympus) towers to 24 km above a surrounding surface, and is topped by a 65-km caldera. Most astronomers believe it to be extinct, but we cannot be sure, and in any case Mars is not a changeless world. There are frequent clouds and dust-storms, and sometimes the dust is so extensive that it covers the entire planet, hiding the surface features completely.

There are, too, features which are almost certainly old riverbeds, and there is evidence of past flash-floods. This shows that Mars used to have much more atmosphere than it has now, and it may well be that life gained a foothold there, dying out when conditions became unsuitable. We will know for certain only when we manage to obtain samples of Martian material for analysis in our laboratories. I admit to being sceptical about the possibility of finding Martian fossils, but they cannot be entirely ruled out.

The map given here shows the main markings on Mars which can be seen with a telescope of, say, 30-cm aperture. Remember, however, that Mars can only be well seen when it is fairly near opposition. At its best, it may outshine every other planet apart from Venus; at its faintest it may easily be mistaken for a fairly bright red star.

Mars has two satellites, discovered as long ago as 1877 and named Phobos and Deimos. Both are very small, and are irregular in shape; Phobos has a longest diameter of 27 km, Deimos only 15 km, so that they are quite unlike our massive Moon, and seem to be ex-asteroids which were captured by Mars long ago. Photographs sent back by the various space-craft show them to be rough and cratered, with very dark surfaces. Neither would be of much use in providing light during the Martian nights, and from regions near the poles they could never be seen at all, because they would remain below the horizon. Phobos behaves in a curious way. It moves round Mars at a distance of only 5879 km above the planet's surface, and has a period of 7 hr 39 m, so that it completes three orbits in every Martian

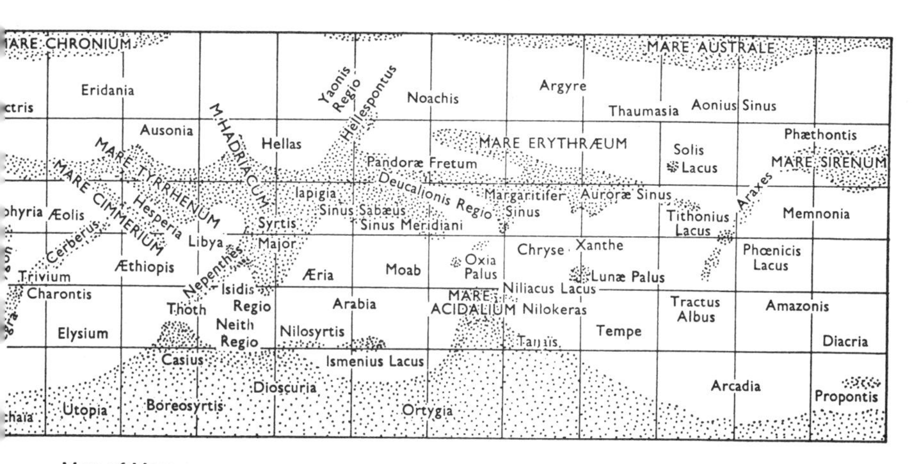

Map of Mars.

day. Seen from the surface, it would rise in the west, gallop across the sky, and set in the east 4½ hr later. Deimos is further out, and would remain above the horizon for two and half sols at a time, but it would give little more light than Venus does to us.

Mars is almost within reach. By the middle of the next century, at the latest, people should have landed there, and they will find a fascinating world awaiting them, even if there are no canal-building Martians!

Chapter 10
The Outer Planets

Passing through the belt of asteroids — of which more anon — we come next to Jupiter, which is the giant of the Sun's family and is more massive than all the other planets put together. We have a long way to go, because Jupiter's mean distance from the Sun is 778 million km, so that a space-craft of the kind we can build today takes a year and a half to complete the journey even if it goes by the shortest route. Jupiter's 'year' is almost twelve times as long as ours. It reaches opposition every year, so that in this respect it is much more co-operative than Mars.

Jupiter is 143,885 km in diameter as measured through its equator, but the polar diameter is only 133,708 km. The globe is very obviously flattened, as even a small telescope will show. This is because Jupiter is spinning round very quickly; a Jovian 'day' is less than 10 hr long, and is not the same all over the planet. At the equator, the rotation period is 9 hr 50½ m. Over the rest of the globe it is about five minutes longer, but various discrete features have periods of their own, so that they drift around in longitude. This sort of behaviour is only to be expected when we are dealing with a purely gaseous surface.

The main markings on Jupiter are the cloud belts. Generally there are several of these visible with modest equipment, though they change in intensity and in form quite quickly. There are also spots, wisps and festoons. The most famous feature is the Great Red Spot, a huge oval with a surface area greater than that of the Earth; it has been seen on and off (more on than off) ever since the seventeenth century, and though it can disappear for a while it always comes back. It lies in Jupiter's southern hemisphere, shifting in longitude but staying at almost the same latitude. It was once thought to be a solid body, but we have now found that it is a whirling storm, coloured possibly by phosphorus rising from below the cloud-deck. Whether it will last indefinitely remains to be seen.

Jupiter seems to have a solid core at a very high temperature (at least 30,000 °C) surrounded by layers of liquid hydrogen, above which comes the 'atmosphere', which is about 1000 km deep and is made up chiefly of hydrogen, together with helium and some unpleasant hydrogen compounds such as ammonia and methane. We do know that Jupiter sends out more energy than it would do if it depended entirely upon what it receives from the Sun, but this is not to say that it qualifies as a 'junior star'. It is not massive enough or hot enough at its core for that, and the outer clouds are bitterly cold, at a temperature of around -150°C.

The four space-craft which have flown past Jupiter (Pioneers 10 and 11, and Voyagers 1 and 2) have confirmed that there is a very powerful magnetic field, together with zones of radiation which would promptly kill any astronaut foolish enough to venture inside them. Jupiter is definitely a place to be viewed from a respectful distance.

As befits the senior member of the Solar System, Jupiter has a large family of moons or satellites; sixteen in all. Twelve of these are very small, but the remaining four are large enough and bright enough to be seen with any telescope, or even good binoculars. They were studied as long ago as 1610 by Galileo, the first great telescopic astronomer, and are known collectively as the Galileans. Their individual names are Io, Europa, Ganymede and Callisto.

Armed with a telescope, the observer can be greatly entertained by following the movements of the Galileans from night to night. Their orbital periods range from 1 day 19 hr for Io out to 16 days 18 hr for Callisto. They may pass in transit across the Jovian disk, together with their shadows, and they may also be occulted by the planet, or eclipsed by Jupiter's shadow. All are substantial. Io is slightly larger than our Moon, Europa slightly smaller, and Ganymede and Callisto much larger. Indeed, the diameter of Ganymede is slightly greater than that of Mercury, though it is not so massive and, like the other Galileans, is bereft of atmosphere.

All four were studied from close range by the Voyagers, in 1979. Ganymede and Callisto are icy and cratered; Europa is also icy, but has a smooth surface criss-crossed with shallow cracks; Io is red and highly volcanic, with sulphur eruptions going on all the time. Photographs of it make it look rather like an Italian pizza. It must be just about the most lethal world in the Solar System, because not only is its surface wildly unstable but it also moves in the thick of the dangerous radiation zones. Jupiter itself is a source of radio waves, and there is a strong electrical current running between Io and the planet.

Jupiter has a thin, dark ring, discovered by the Voyagers and unobservable from Earth, but of course the supreme system of rings belongs to the next giant planet, Saturn, which moves round the Sun at a mean distance of 1427 million km in a period of 29½ years. It is smaller than Jupiter, but even so its equatorial diameter is 120,536 km. It too is markedly flattened, and it has a short day, only 10¼ hr long. Its make-up is not unlike that of Jupiter, but it is much less massive, and the mean density of the globe is actually less than that of water. It has been said that if you could put Saturn into a vast ocean, it would float.

Belts can be seen across the yellow disk; spots are rare, though occasional white spots become prominent for a few weeks (as in 1933 and again in 1990).* However, it is the rings which make Saturn so glorious. There are three main rings, two bright and one semi-transparent; when well displayed, the view in even a small telescope is superb. To my mind, at least, Saturn is much the most beautiful object in the entire sky.

SATURN, drawn on 3 May 1983 (10h 10m) by Patrick Moore; 61-cm reflector, Mount John Observatory, x 300.

The rings may look solid, but no solid or liquid ring could possibly exist, because it would promptly be torn to pieces by Saturn's strong gravitational pull. The rings are in fact made up of icy particles, all spinning round the planet in the manner of dwarf moons. The two main rings are separated by a gap known as Cassini's Division, in honour of the Italian astronomer who discovered it as long ago as 1675; there is also a narrower gap, Encke's, in the outer bright ring.

Nobody is sure about the origin of the rings. They may be the remnant of a former satellite which wandered too close to Saturn and paid the supreme penalty, or they may be simply debris which never condensed into a larger body.

The rings measure 270,000 km from one side to the other, but they are less than a km thick, so that when placed edgewise-on to us, as in 1995, they almost disappear. At other times the details are very evident. The drawing shown here was made with the 61-cm at Mount John Observatory at the time when conditions for observation were very favourable.

Several other fainter rings were discovered by the Voyagers, which by-passed Saturn in 1980 and 1981, and detailed studies

* The 1933 spot was discovered by a skilled amateur astronomer named W.T. Hay. You may remember him better as Will Hay, the stage and screen comedian!

were also made of the satellites. Unlike Jupiter, Saturn has only one large attendant, Titan, but there are several of moderate size. Rhea and Iapetus are around 1500 km in diameter, Dione and Tethys around 1100 km, and Mimas, Enceladus and Hyperion above 300 km.

Iapetus is interesting, because it has one hemisphere which is as bright as snow and the other which is as black as a blackboard. Apparently the dark patch is a 'stain' on an icy globe, but its origin is unknown. However, the most important satellite is Titan, which is nearly as large as Mercury and has an escape velocity of 2.5 km per second. It has a thick atmosphere, made up chiefly of nitrogen; there are orange clouds, and all the ingredients for life are there, though the intense cold (around -180°C) means that life has probably been unable to gain a foothold. The Voyager pictures show only the top of a layer of orange 'smog', and we do not know what lies underneath. There may be seas — not of water, but of chemicals such as methane or ethane. In any case, Titan is unlike any other world in the Solar System. We may know more in the year 2004, when a new space-craft is scheduled to go there and make a controlled descent through the clouds.

In 1781 William Herschel, then an unknown amateur astronomer living in Bath in western England, was carrying out what he called a 'review of the heavens', using a home-made reflecting telescope, when he came across an object which did not look like a star. It showed a disk, which no star can do, and it moved slowly from one night to the next. Herschel mistook it for a comet, but before long it was found to be a new planet, moving far beyond the orbit of Saturn. It was named Uranus, after the first ruler of Olympus.

Uranus can just be seen with the naked eye if you know where to look for it, but no Earth-based telescope will show much on its pale, greenish disk. Though it ranks as a giant — its diameter is 51,118 km — it is much smaller than Jupiter or Saturn, and it is 'only' 14 times more massive than the Earth. Its mean distance from the Sun is 2870 million km, twice that of Saturn, and its orbital period is 84 years. Its 'day' is short, only a little over 17 hr in length.

Like its larger brothers, Uranus has a gaseous surface, but its internal make-up is rather different. There is no pronounced source of inner heat, and round the rather ill-defined core there are layers of 'ices', mainly water, ammonia and methane, which would be frozen at the low temperature of the uppermost clouds (around -214°C). But the most curious feature of Uranus is the tilt of its axis.

The Earth's axis of rotation is inclined by 23½° to the perpendicular to the orbit, and this is why we have our seasons.

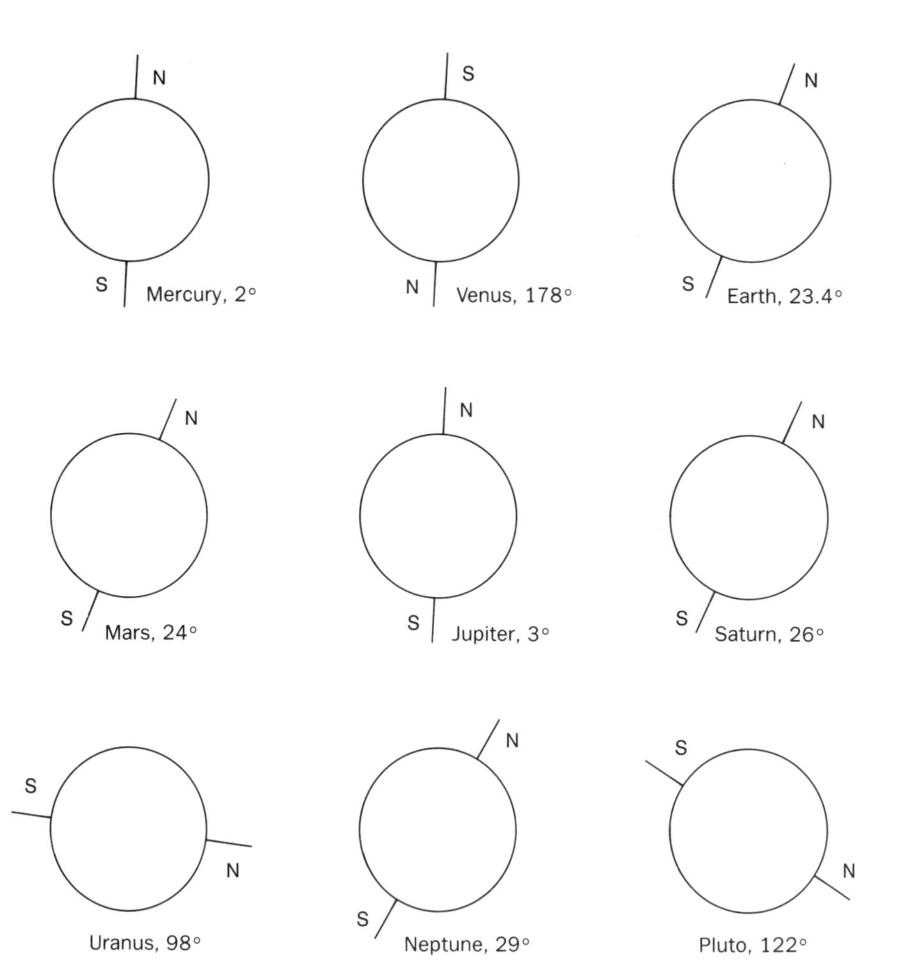

Axial inclinations of the planets.

Most of the other planets have similar tilts, though Jupiter spins almost 'upright'. In the case of Uranus the tilt is 98°, more than a right angle, so that there are times when the pole of the planet lies in the centre of the disk as seen from the Earth, with the equator extending all the way round the limb. This was the situation in 1986, when Voyager 2 made its pass.

Why does Uranus have this strange tilt? We have to confess that we do not know. The favoured idea is that in its early history it was struck by a large body and literally knocked sideways. This does not sound at all plausible, but it is difficult to think of anything better.

Few clouds were recorded by Voyager — Uranus is a remarkably bland planet — but a magnetic field was detected. The magnetic axis is nowhere near the axis of rotation, and does not even pass through the centre of the globe, so that the magnetic poles are a long way from the poles of rotation. It is rather as if the Earth's south magnetic pole lay somewhere in Egypt!

Uranus has a system of thin, dark rings, difficult to see from Earth. There is a large satellite family, even though the largest member of the family (Titania) is only 1578 km in diameter. Before the Voyager mission, five satellites were known; Voyager added another ten, though all the new discoveries are so close to the planet that they cannot be seen from Earth. The five largest satellites (Ariel, Umbriel, Titania, Oberon and Miranda) are icy and cratered; Ariel shows broad, branching valleys, while Umbriel has a dark, subdued surface and Miranda has a very complex appearance, with craters, ice-cliffs and large regular features which have been likened to race-tracks!

A few years after Uranus had been discovered, mathematicians found that it was not moving in the way that had been expected. Some unknown force was pulling it out of position. The cause of the trouble was thought to be a new planet, moving round the Sun at a much greater distance, and in 1856 the planet was tracked down. The actual discovery was made from the Berlin Observatory by Johann Galle and Heinrich D'Arrest, based on calculations by the French astronomer U.J.J. Le Verrier, though it must be added that an Englishman, John Couch Adams, had made similar calculations slightly earlier and had reached virtually the same result.

Neptune, as the planet was named, is almost a twin of Uranus. It is very slightly smaller, but considerably more massive; its disk is bluish rather than green, and it is too faint to be seen with the naked eye, though a small telescope will show it. The revolution period is 164.8 years, so that we have known it for less than one Neptunian year; the 'day' is 16 hr long.

Uranus and Neptune may be twins, but they are not identical. Neptune has a much hotter core, and does not share in Uranus' extraordinary axial tilt, though, as with Uranus, the magnetic axis is well away from the axis of rotation, and the magnetic field itself is weaker. It is a much more dynamic world than Uranus, as was found when Voyager 2 sped past it in 1989. There is one major feature, the Great Dark Spot, together with other spots, and a variable marking which was nicknamed the Scooter because it has a relatively quick rotation period (less than 16 hr). Windspeeds are high, and there are methane clouds hanging high above the main cloud-deck. There is a system of dark rings, but they are much less obvious than those of the other giant planets; and, predictably, Neptune is a source of radio waves.

Two satellites were known in pre-Voyager days; one large (Triton) and one small (Nereid). Both were unusual, though in different ways. Triton, rather smaller than our Moon, moves round Neptune in a retrograde or wrong-way direction — that is to say, in a sense, opposite to that of Neptune's rotation. It is the only major retrograde satellite in the Solar System, and it seems likely that it used to be an independent body which was captured by Neptune in the remote past. Voyager 2 showed that the visible pole was covered with pink nitrogen snow, and that there are active geysers. Apparently there is a layer of liquid nitrogen below the surface; if this migrates upward for any reason, the pressure is relaxed, and the result is an outrush of nitrogen ice and vapour. Nothing of the sort had been expected, particularly as Triton, with a temperature of -235°C, is the coldest world ever encountered by a space-craft. There is little surface relief — there are no lofty mountains or deep craters on Triton — and there is a thin but detectable atmosphere.

Nereid is much smaller (around 240 km in diameter) and has a very eccentric orbit, more like that of a comet than a satellite. It moves round the planet in a period of almost a year, and the distance between it and Neptune ranges between 9,688,500 km and only 1,345,500 km , so that it is always much further out than Triton. Voyager added six new satellites, all very close-in and therefore unobservable from Earth, though one of them (Proteus) is actually larger than Nereid.

The ninth planet, Pluto, was discovered in 1930 by Clyde Tombaugh from the Lowell Observatory at Flagstaff, in Arizona. Its position had been predicted fairly accurately — by Percival Lowell himself — from irregularities in the movements of Uranus and Neptune, but as soon as it was found it began to pose real problems for theorists. It is faint, so a telescope of some size is needed to show it (I can see it with my 32-cm reflector, and no doubt keener-eyed observers can do better), and its diameter is less than that of the Moon, or even Triton. It is of very slight mass and seems to be made up of a mixture of rock and ice. It cannot possibly have any measurable effect upon the movements of giant planets — one might as well try to divert a charging hippopotamus by throwing a baked bean at it — and therefore it cannot be the planet which was being sought. In this case, the real 'Planet X' may await discovery.

Pluto has a curious orbit. Its revolution period is 248 years, and its path is much less circular than those of the other planets, so that for part of the time, as at present, it is closer-in than Neptune. Not until 1999 will it again become 'the outermost planet'. It is not a solitary traveller, because it is accompanied by another body, Charon, which has more than half the diameter of Pluto itself.

On the whole it does not look as though Pluto can be ranked as a proper planet. It is too small and too low in mass; it has the wrong sort of orbit, and it may be only the brightest member of a whole swarm of similar objects moving round the Sun well beyond the path of Neptune. All we can say about it is that it has an icy surface, with a very thin atmosphere which may well freeze out when Pluto moves away to the further part of its orbit and becomes even colder than it is now. Unfortunately, no spacecraft has gone anywhere near it.

I appreciate that this is a very sketchy description of the main members of the Sun's family, but I hope that it is enough to show that all the planets have their own points of special interest. Armed with even a modest telescope, you can have endless enjoyment by looking at the phases of Venus, the changing ice-caps of Mars, the belts and spots of Jupiter, the celestial hide-and-seek played by the Galilean satellites, and the glorious ring-system of Saturn. Try for yourself, and I am sure that you will agree with me.

Chapter 11

Cosmic Debris

Before leaving the Solar System, we must say at least something about its minor members. Insubstantial though they may be, they are of great interest, and are much more important than originally thought.

The asteroids, or minor planets, swarm in the wide gap between the orbits of Mars and Jupiter. Only one (Vesta) is ever visible with the naked eye, and then as nothing more than a dim speck. Only one (Ceres) is as much as 900 km in diameter, while most of the rest are real midgets. The total membership is probably several tens of thousands. By 1993 more than 5000 asteroids had had their paths worked out.

The first asteroid, Ceres, was discovered on 1 January 1801, the first day of the new century. Three more (Pallas, Juno and Vesta) followed by 1808; No.5, Astraea, was tracked down in 1845, and since 1847 no year has passed by without more being added to the list. Initially it cannot be said that they were popular members of the Sun's family, because photographic plates exposed for quite different reasons were often found to be crawling with asteroid tracks, all of which had to be identified and eliminated and all of which wasted an incredible amount of time. One irritated German astronomer referred to the 'Kleineplanetenplage' or Minor Planet Pest, and another dismissed them as 'vermin of the skies'. Only in our own time have they come to be regarded as respectable and welcome members of the Solar System!

No asteroid, even Ceres, is massive enough to retain a trace of atmosphere. Some members of the swarm seem to be stony in composition, while others are metal-rich; some are fairly reflective, others very dark. The idea that they represent the debris of an old planet which broke up is not now favoured, and it seems more likely that no large planet could ever form in this region because of the presence of Jupiter. Every time a planet started to form, Jupiter's powerful pull tore it apart, so that the end product was a shoal of dwarfs.

Asteroid names are a source of interest and, sometimes, hilarity. Mythological names soon ran out, and some very curious characters have found their way into the sky — notably Mr. Spock, named not directly after the sharp-eared Vulcanian of *Star Trek*, but after a ginger cat which was itself named after

Mr. Spock. I cannot resist adding that Asteroid 2602, discovered by Dr Edward Bowell from America, has been officially named 'Moore' after me. It is about 3 km in diameter, and moves near the outer edge of the main asteroid belt.

The first close-range picture of an asteroid was taken on 13 November 1991 from the Galileo space-craft, which was on its way to Jupiter and was in the thick of the main belt. The target asteroid, Gaspra, was imaged from 16,000 km, and proved to be wedge-shaped, not unlike a distorted potato, with a darkish, rocky, crater-pitted surface. The longest diameter is only 20 km, and it seems certain that Gaspra is only a part of a larger body which broke up. After all, collisions between asteroids must be fairly frequent. Galileo subsequently imaged another asteroid, 243 Ida, which is somewhat larger, measuring 56 × 24 × 24 km, though it is like Gaspra in being irregular and cratered. Ida has also been found to have a 1.5 km satellite, which showed up well in the Galileo picture. Another asteroid, Toutatis, has been found to be double, with two components which are virtually in contact; a radar image shows it well.

Some asteroids swing away from the main swarm, and may pass close to the Earth. There are more of them than we used to think, but all are tiny. One of them, not yet named and known merely by its temporary designation of 1993 KA, is thought to be no more than 11 m across, and may be even smaller than this. In May 1993 it brushed past us at a mere 140,000 km, roughly one-third the distance of the Moon. It has an eccentric path; at its closest to the Sun it moves inside the orbit of Venus, while at aphelion it is almost as far away as the orbit of Jupiter.

Is there a chance that we will be struck by one of these wandering asteroids? The answer must be 'yes', and undoubtedly it has happened in the past. There is even a well-supported theory that an impact about 65 million years ago caused such a change in our climate that the dinosaurs, which had been lords of the world for so long, could not adapt to the new conditions, and died out. I admit to being somewhat sceptical, but at least it is a possibility. If it happens again, however, no doubt we will be able to cope with the crisis better than the dinosaurs did.

There are also asteroids well beyond the main swarm. The so-called Trojans move in the same orbit as Jupiter, though they keep prudently either well ahead of or well behind the Giant Planet, and are in no danger of being swallowed up. No.2060, Chiron, spends most of its time between the orbits of Saturn and Uranus. It is probably at least 150 km across, which is large by asteroidal standards; it has a revolution period of 50 years, and when closest to the Sun it develops what appears to be a temporary atmosphere, so that presumably there are surface ices which start to evaporate when warmed. Recently (1992-3)

American astronomers have located asteroidal objects which are even further out, and move well beyond the orbit of Neptune. They seem to be around the same size as Chiron, and it is quite likely that many more of them await discovery, in which case the true planetary system is rather more extensive than we used to think.

Then, of course, there are comets. To recapitulate: a comet is made up essentially of an icy nucleus, mixed with 'rubble', never more than a few kilometres across. When the comet nears the Sun, the ices begin to evaporate; the comet develops a head or coma, and very often a tail or tails, all of which disappear when the comet returns to the cold regions in the far part of the Solar System.

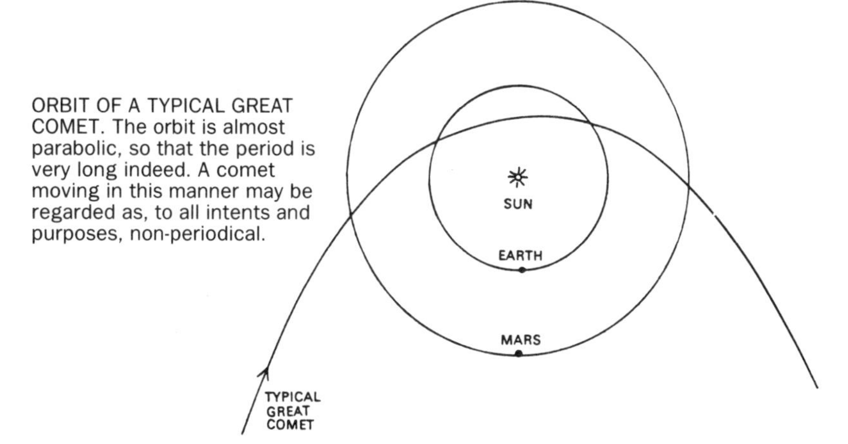

ORBIT OF A TYPICAL GREAT COMET. The orbit is almost parabolic, so that the period is very long indeed. A comet moving in this manner may be regarded as, to all intents and purposes, non-periodical.

Most comets move round the Sun in orbits which are very elliptical. Short-period comets take only a few years to make a full journey round the Sun. Encke's Comet has a period of only 3.3 years, and even at aphelion it does not venture out as far as the orbit of Jupiter.* Other comets have longer periods: 76 years for Halley's Comet, 130 years for Swift-Tuttle, and so on. Then there are the very bright comets whose periods amount to hundreds, thousands or even millions of years, so that they cannot be predicted, and are always liable to arrive unheralded and take us by surprise. It is not generally thought that comets are true members of the Solar System, and were formed from the original cloud of material associated with the young Sun; they do not come from interstellar space. According to the

* Comets are generally named after their discoverers, but sometimes after the mathematician who first computed the orbit. Encke's Comet was first seen in 1786 by Pierre Méchain, but its periodicity was first recognised in 1822 by J.F. Encke. Sometimes there is more than one discoverer; thus Comet Swift-Tuttle was found independently, in 1862, by Lewis Swift and Horace Tuttle.

Orbit of Encke's Comet; period 3.3 years.

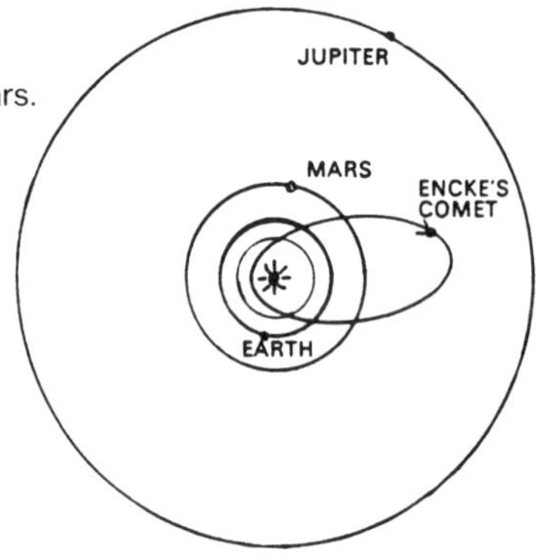

Dutch astronomer Jan Oort, there is a whole 'cloud' of cometary bodies orbiting the Sun at a distance of between one and two light-years. If one of these bodies is perturbed for any reason, perhaps by a passing star or an unknown planet, it will start to fall inward toward the Sun, and after a journey which takes an extremely long time it invades the inner Solar System, where we can see it. One of several things may happen. The comet may simply swing past the Sun and return to the Oort Cloud; it may be so perturbed by a planet, usually Jupiter, that it is thrown out of the Solar System altogether; it may be captured by a planet and forced into a short-period orbit; or it may fall into the Sun and be destroyed. One comet, Shoemaker-Levy, was captured by Jupiter in 1992 and actually hit the planet during the northern hemisphere summer of 1994. A comet, remember, is of very slight mass, so that it is very easily pulled around. I have described a comet as being 'the nearest approach to nothing that can still be anything'.

There are two widespread misconceptions about comets. First, no comet flashes quickly across the sky; if you see an object which is moving perceptibly against the starry background, it certainly cannot be a comet, and must be either a meteor or an artificial satellite (unless of course it is something much more mundane, such as an aircraft). Because a comet is far beyond the Earth's atmosphere and is millions of kilometres away, its motion against the stars is very slow, and one has to watch it for hours to note that it is shifting at all. Secondly, not all comets have tails. Brilliant visitors look imposing; for example Bennett's Comet of 1970, discovered by the South African astronomer of that name, made a brave showing for several weeks. On the other hand many short-period comets have no tails at any time, and appear only as dim misty patches in the sky, so that it is only too easy to confuse them with faint star-clusters or nebulae. Not many of them reach naked-eye visibility.

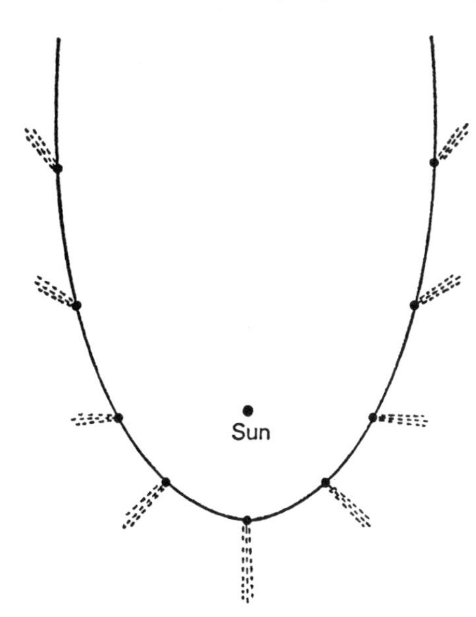

DIRECTION OF A COMET'S TAIL. The tail always points more or less away from the Sun, so that when travelling outward the comet moves tail-first. At great distances, the tail disappears completely.

There are two types of tails: dust tails and gas tails. Both are formed from material expelled from the nucleus, so that a comet loses a certain amount of mass every time it passes near the Sun, and is bound to waste away. Indeed, several short-period comets seen regularly in past years have now disappeared — such as the famous Biela's Comet, which broke in two in 1846 and has not been seen, as a comet, since 1852, though for some time we used to see its debris in the form of meteors.

Tails always point away from the Sun, so that when a comet is travelling outward it actually moves tail-first. The dust tail is due to the repelling effect of the solar wind, while the gas or ion tail is due to the pressure of sunlight — and light does exert a pressure, albeit a very tiny one. Some comets develop tails of both types.

Incidentally, it has been suggested that some of the tiny close-approach asteroids may simply be ex-comets which have lost all their volatiles. The classic case is that of Asteroid 4015, which was discovered some years ago and was subsequently found to be identical with Comet Wilson-Harrington of 1949. There can be no doubt that the object had a tail then, and is now purely asteroidal in appearance. I had always been sceptical about the proposed link between asteroids and comets, but I have now started to have second thoughts.

Really 'great' comets can be magnificent — so I am told; I have never seen one, because they have been very scarce of late. During the last century they were fairly frequent. For example, there were brilliant comets in 1811, 1843, 1858, 1861 and 1882, all of which were visible with the naked eye in broad daylight and most of which cast shadows. Unfortunately the last great comet was seen as long ago as 1910, and we cannot tell when there will be another, though astronomers live in hope.

Only one periodical comet can ever become really bright. This is Halley's, named after the second English Astronomer Royal, who saw it in 1682 and successfully predicted its return in 1758. It comes back every 76 years or so (the period is not absolutely constant) and has been seen regularly since before the time of Christ. The last three returns have been those of 1835, 1910 and 1986. On the first two occasions it was imposing, but in 1986 it was badly placed. However, five space-craft were sent to it, and one, the British-built Giotto, went right into the head, sending back close-range pictures of the nucleus, which was found to be coated with warm, dark dust and to have active jets. At each return to the Sun it is estimated that Halley's Comet loses 300 million tons of material, and it is not really large, as the nucleus measures only 15 x 8 x 8 km. It will last for some centuries yet before wasting away, and we can be sure that it will return in 2061, though it will again be badly placed and will be best seen from the Earth's northern hemisphere. Do not be discouraged; I can assure you that at the return of 2137 it will be striking!

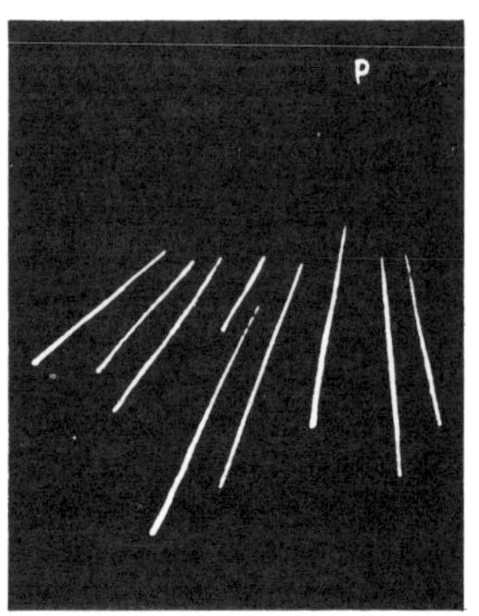

METEOR RADIANT. The meteors in a shower are moving through space in parallel paths. As seen from Earth, they will seem to diverge from the radiant point (P).

As a comet moves along, it leaves a dusty trail behind it. When the Earth ploughs through one of these trails, we collect a large number of particles, and the result is a shower of meteors or shooting-stars. Because the meteors in any shower are moving through space in parallel paths, they seem to come from the same point in the sky, which is termed the radiant of the shower. The effect is well shown by a pair of parallel roads, which will seem to meet at a point near the horizon.

There are many annual showers. Pride of place must go to the meteors of early August, which radiate from the constellation Perseus and are therefore known as the Perseids.

The shower can be very rich, although the radiant is in the northern hemisphere of the sky and the display is better seen from Europe than from the south; from Invercargill, Mirphak, the brightest star in Perseus, never rises at all. Of course, the meteors are not confined to the Perseus area; they spread widely across the sky, and are active between around July 27 and August 18, with a maximum in August 11-12. The parent comet is Swift-Tuttle, which has a period of 130 years. It last returned in 1992, and we expected a really rich shower of Perseids, though in fact it was not much more imposing than usual — meteors, like politicians, are notoriously unreliable!

The November meteors come from a radiant in Leo, the Lion, which also is in the northern part of the sky, though not so far from the equator. The parent comet of the Leonids is Tempel-Tuttle, which has a period of 33 years. Unlike the Perseids, the Leonids are very erratic; sometimes they are very sparse, occasionally they can produce 'meteor storms', when shooting-stars rain down like snowflakes. This happened in 1799, 1833, 1866 and 1966. There is every prospect of another impressive display of cosmic fireworks in 1999, so be on the alert!

There are many minor showers, and there are also sporadic meteors, not associated with known comets, which appear from any direction at any moment. What we see, of course, is not the tiny particle itself, but the luminous effects in the air. A meteor dashes into the atmosphere at a speed of anything up to 72 km per second, so that it sets up a great deal of friction, but it burns away before it has penetrated to within 60 km of the ground.

Larger bodies which enter the atmosphere can survive the whole drop to the Earth's surface before being burned away, and may even produce craters. These are termed meteorites. Note that a meteorite is not simply a large meteor, and there is no link between the two classes of objects. Meteorites come from the asteroid belt, and are not connected with comets. Indeed, there seems to be no difference between a large meteorite and a small asteroid; it is simply a question of terminology.

Some meteorites are made up chiefly of iron, while others are stony. Most are small, and museums have collections of them, but really large specimens are known. The holder of the 'heavyweight record' is still lying where it fell in prehistoric times, at Hoba West, near Grootfontein in Southern Africa. It weighs over 60 tons, and has lain undisturbed ever since it landed — though, alas, United Nations soldiers stationed in the area quite recently started to vandalize it before they were firmly ordered to desist.

Meteorites have fallen in every part of the world, but obviously they are most easily found in areas which are sparsely populated, for example Western Australia and, in particular, Antarctica. The largest known Australian specimen is the Mundrabilla Meteorite, which has a weight of twelve tons.

Impact craters are found here and there. In 1902 a missile landed in Siberia and blew pine-trees flat over a wide area, but without producing a crater. Another large meteorite fell near Vladivostok in 1947, but broke up before landing, and produced a whole crop of small craters. The most famous impact crater is in Arizona; the diameter is 1265 metres, and it is well-formed. The age is certainly well over 20,000 years, and may be more like 50,000 years (opinions differ). It has become very much of a tourist attraction, and is easily reached, not far off Highway 99.

However, you need not go as far as America. No impact craters have been identified in New Zealand, but there are several in Australia. At Henbury, in the Northern Territory, there is a whole cluster of craters, no doubt produced by the same object which broke up before impact; another crater, largely filled with woodland, is not far off, at Boxhole. The best example is Wolf Creek, just over the border of Western Australia. It is not too easy to reach, but the road from the nearest township, Halls Creek, is usually open from October to May. It was discovered from an aerial survey in 1947, and by geological standards it is young; its age cannot be more than 15 million years, and 2 million years is a more likely estimate. There is no problem in scrambling up the wall and looking down into the crater, though descending to the floor takes much longer, and the flies are a constant nuisance.

There is a legend about it. The local Djaru Aborigines call it Kandimalal, and they describe two rainbow snakes whose sinuous path across the desert formed Wolf Creek and the adjacent Sturt Creek; the crater, they say, marks the point where one of the snakes emerged from below ground. It is also worth visiting Gosse's Bluff, in the Northern Territory, where the crater is much older and is largely eroded.

The Vredefort Ring lies near Pretoria, in South Africa. Many lists class it as an impact structure, but geologists who have made long-term studies of it, and know every facet of it, are unanimous in saying that it is a volcanic feature and is not of cosmic origin. Inside it are two small towns; Parys, and Vredefort itself. A larger, mainly water-filled impact structure has recently been identified near Pretoria.

If a meteorite hit a city or a densely-populated area, the death-roll would be considerable, but fortunately the chances of anything of the sort are low, and there is no authentic record of anyone having been killed by a meteorite, though it is true that a few people have had narrow escapes. These include two American boys, Brodie Spaulding and Brian Kinzie, on 31 August

1991, when a meteorite landed less than 4 m away from them and made a crater 9 cm wide.

Tektites are small, glassy objects found only in a few localized regions, one of which is in Australia. They seem to have been heated twice, and are aerodynamically shaped. They were once thought to be unusual types of meteorites, but most authorities now believe that they are of volcanic origin and do not come from the sky.

The space between the planets is not empty, and there is always a considerable amount of thinly-spread material, much of it lying in the main plane of the Solar System. When illuminated by the Sun, it shows up as a cone-shaped glow known as the Zodiacal Light. It extends away from the Sun, and is generally observable for only a fairly short period after the Sun has disappeared or before it rises. The best months to see it are February/March and September/October. You need a very clear sky, well away from artificial lights, but at its best the Zodiacal Light may be as bright as the Milky Way. Finally there is the Gegenschein or Counterglow, also due to interplanetary debris, which appears as a faint patch of light exactly opposite to the Sun in the sky. It is very elusive. I admit that I have seen it only once in my life — in 1942, from an England blacked out as a precaution against German air-raids — and so far as I know, it has never been properly photographed.

So much for our tour of the Solar System. There is plenty to see, and plenty to discuss, but let us now turn to the wide universe, and move out into the realm of the stars.

Chapter 12
The Stellar Universe

Look up into the sky on a dark, clear night, and you may think that you can see millions of stars. Yet as is so often the case, appearances are misleading. If you can see as many as 3000 stars without using binoculars or a telescope, you are doing very well indeed. The stars are not all alike; some are brilliant and others dim, while there are also marked differences in colour. In the Southern Cross, for example, three of the stars in the main pattern are bluish-white, while the fourth is orange-red.

A star's brightness is given by its apparent magnitude. The scale works in the manner of a golfer's handicap, with the more brilliant performers having the lower values; thus a star of magnitude 1 is brighter than a star of magnitude 2, 2 is brighter than 3, and so on. A few stars are brighter than the first magnitude, and have zero or negative values. The dimmest stars normally visible with the naked eye on a clear night are of magnitude 6; a 7.6-cm telescope will take you down to almost magnitude 10, and with my 38-cm reflector I can reach magnitude 14 without much difficulty. Incidentally, the apparent magnitude of the sun is almost minus 27.

The constellation patterns — which, as we have noted, mean nothing at all, because the stars are at very different distances from us — were introduced long ago. Generally we use the Latin names, and I propose to do so here, because there is nothing in the least difficult about them. The brightest stars have individual names, most of which were given by the Arab astronomers of 1000 years ago, but in general these names are used only for very bright stars, plus a few others of special interest. Otherwise, we simply use catalogue numbers.

In 1603 a German amateur, Johann Bayer, drew up a star catalogue in which he allotted Greek letters to the stars in each constellation, beginning with the brightest star, Alpha, and working through to Omega. In case you are not familiar with the Greek alphabet, it is on the opposite page.

The sequence is not always followed. For instance in Sagittarius, the Archer, the brightest stars are Epsilon and Sigma, with Alpha and Beta very much 'also rans'. Still, the system is reasonably convenient, and it has stood the test of time. It works well for Crux, the Southern Cross, where the four brightest stars are Alpha (mag. 0.8), Beta (1.2), Gamma (1.6) and Delta (2.8).

α Alpha
β Beta
γ Gamma
δ Delta
ϵ Epsilon
ζ Zeta
η Eta
θ Theta
ι Iota
κ Kappa
λ Lambda
μ Mu
ν Nu
ξ Xi
ο Omicron
π Pi
ϱ Rho
ς Sigma
τ Tau
υ Upsilon
ϕ Phi
χ Chi
ψ Psi
ω Omega

In addition, stars have numbers which were allotted by the first English Astronomer Royal, John Flamsteed, though this applies only to the stars which rise over the British Isles. Thus Sirius in Canis Major, the Great Dog, is also known as Alpha Canis Majoris and as 9 Canis Majoris.

The fact that we live in the Solar System makes it rather hard for us to appreciate how unimportant we are. The Sun is a normal star, but cosmically speaking it is right on our doorstep, and when we come to consider the distances of the other stars we are, so to speak, batting on an entirely different wicket.

No telescope will show a star as anything but a dot of light. (If you look at a star and see it as a large, shimmering disk, you may be sure that there is something wrong with the optics, or that the telescope is completely out of focus.) Therefore, distance-measuring is not easy, even with the nearest of all the brilliant stars — Alpha Centauri,* the brighter of the two Pointers to the Southern Cross, which is just over four light-years away. The first successful attempt to measure its distance was made in the 1830s by Thomas Henderson, during his brief spell as Director of the Cape Observatory.

Henderson selected Alpha Centauri partly because it has a slight but definite individual or proper motion against the background of more remote stars, and partly because it is a wide, easy double. Henderson decided to attack the problem by using the method of parallax, the principle of which is easy to understand, and which can be demonstrated by a simple experiment.

Close one eye, hold up your finger, and align it with some relatively distant object, such as a tree (see diagram). Now, without moving, close your first eye and open the other. Your finger will no longer be lined up with the tree, because you are observing it from a slightly different direction. If you know the distance between your eyes, and if you can measure the angle

* Curiously, there are no universally accepted names for some of the brightest far-southern stars. Alpha Centauri has been called Toliman, Rigel Kent and Rigel Kentaurus; Beta Centauri has been known as Agena or Hadar; Alpha Crucis as Acrux and Beta Crucis as Mimosa, but the names are not always used as they are in the cases of the other really brilliant stars, such as Sirius and Canopus.

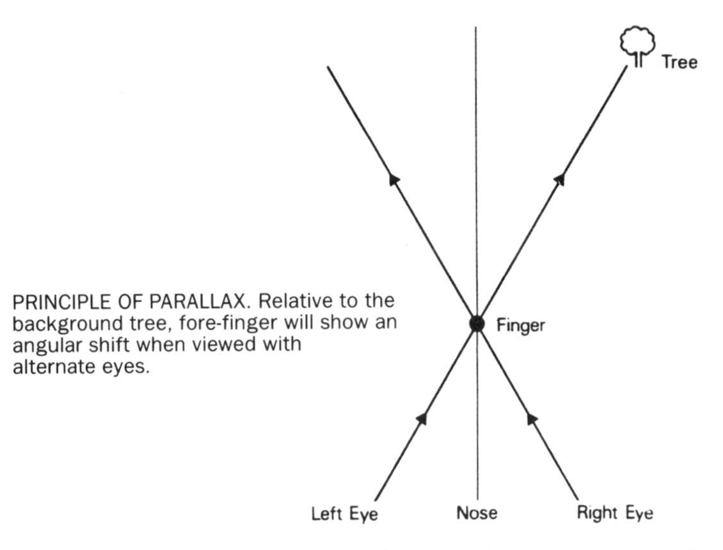

PRINCIPLE OF PARALLAX. Relative to the background tree, fore-finger will show an angular shift when viewed with alternate eyes.

by which your finger seems to have shifted, you can work out the distance between your finger and your face. This is, basically, the method used by a surveyor who wants to measure the distance of some inaccessible object, such as a mountain-top; he observes the object from the opposite ends of a known base-line.

Henderson wanted a very long base-line indeed, so he chose the orbit of the Earth round the Sun. In the next diagram, which is not to scale, the Earth is shown in two positions: in December (D) and in June (J). During the interval, the Earth has shifted from one side of its orbit to the other. Since the Earth is 150 million km from the Sun, the distance between D and J is twice 150 million = 300 million km. The Sun itself is shown at S.

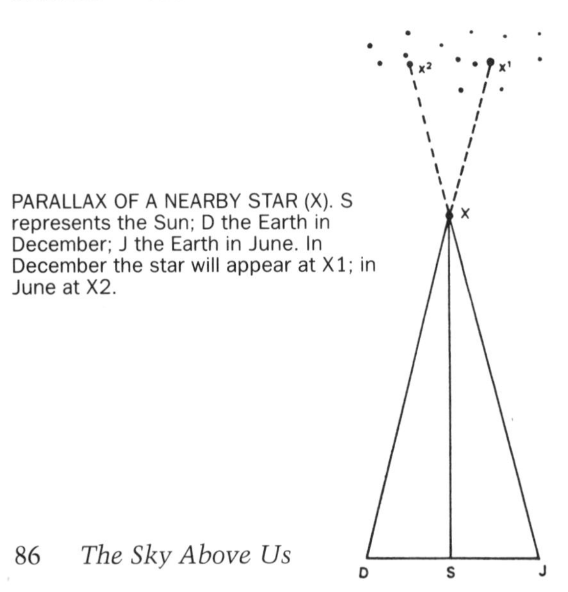

PARALLAX OF A NEARBY STAR (X). S represents the Sun; D the Earth in December; J the Earth in June. In December the star will appear at X1; in June at X2.

When the Earth is at D, our nearby star (X) will be seen at X1 relative to the more remote stars in the background. When the Earth is at J, the star will be seen at X2. The angular shift can be measured, and we now have enough information to calculate the length of the line XS, which is, of course, the distance of the star.

Henderson applied this method to Alpha Centauri, and arrived at a value for XS of just over four light-years, or roughly 40 million million km. He was wise in his selection, since Alpha Centauri is much the closest of all the bright stars, and its faint companion Proxima — a red dwarf, much too dim to be seen without a powerful telescope — is the nearest star of all; its distance is 4.225 light-years. Even so, the parallax shift amounts to less than one second of arc, which needs very careful measuring indeed.

Unfortunately, Henderson did not hurry in working out his results and publishing them. He was not happy at the Cape, and he was only too anxious to return to his beloved Scotland as soon as possible, so that by the time he had finished his calculations he had been forestalled. In Germany, Friedrich Wilhelm Bessel had made similar parallax measurements of a faint star in the Swan, 61 Cygni, and had derived a distance for it of just over 11 light-years. To Bessel, then, goes the official honour of being the first man to measure the distance of a star. At about the same time another astronomer had been working on the problem — F.G.W. Struve, in Russia. His target was the brilliant blue northern star Vega, and his result was of the right order, though it was less accurate than those of Henderson or Bessel. Vega is 26 light-years from us, so that the parallax shift is much smaller.

The parallax method is suitable for the nearest stars, but beyond a few hundred light-years the shifts become so small that they are swamped by inevitable errors of observation, and other means have to be found. Basically, these depend upon finding out how luminous the star really is. Once this is known, the distance can be worked out — bearing in mind that there are many complications to be taken into account, such as the absorption of light in space.

Here we come straight on to spectroscopy, without which the astronomer would be hopelessly handicapped. I have already said something about the spectrum of the Sun, which, as you will remember, is a rainbow background crossed by dark lines. Each line is characteristic of some particular element or group of elements, and exactly the same principles apply to the spectra of the stars.

With the Sun there is plenty of light to spare, but studying the spectra of faint stars is much more of a problem, and involves using spectroscopic equipment combined with powerful telescopes. The first major surveys were made in the latter part of the nineteenth century, and it was found that the stars could be divided into various quite well-defined spectral types. Today these types are denoted by letters of the alphabet, and can be summed up quite concisely:

Type	*Colour*	*Surface temperature °C*	*Examples*
W	White	80,000	Regor
O	White	40,000 — 35,000	
B	Bluish-white	25,000 — 12,000	Rigel
A	White	10,000 — 8000	Sirius
F	Yellowish	7500 — 6000	Canopus, Procyon
G	Yellow giants	5500 — 4200	Capella
	dwarfs	6000 — 5000	The Sun
K	Orange giants	4000 — 3000	Aldebaran
	dwarfs	5000 — 4000	
M	Orange-red giants	3400	Betelgeux
	dwarfs	3000	Proxima
R	Reddish	2600	
N	Reddish	2500	
S	Red	2600	Chi Cygni

(R and N are sometimes combined as Type C)

You may wonder why the sequence, from hot to cool stars, is alphabetically chaotic. The reason is that there were many errors in the early systems, and some types, such as D and E, were found to be unnecessary. But there are other points to note, too. Stars of types G, K and M are divided into giants, which are very large and powerful, and dwarfs, which are small and dim. This was first established in the early part of our own century by two astronomers, Ejnar Hertzsprung of Denmark and Henry Norris Russell of America, who drew up what are now called Hertzsprung-Russell or HR Diagrams.

In an HR Diagram, the stars are plotted according to their real luminosities, which can usually be found from studying their spectra, and their spectral types. A glance at the HR Diagram given here shows that there is nothing really random about it. Most of the stars lie on a line from the top left to the bottom right; this is known as the Main Sequence, and the Sun is a typical Main Sequence star. Reddish stars are divided into giants and dwarfs. There are no red stars equal in luminosity to the Sun, for example. (*En passant*, I am here considering only the types from B to M. Stars of types W, O, R, N and S are rare, and need not concern us for the moment.)

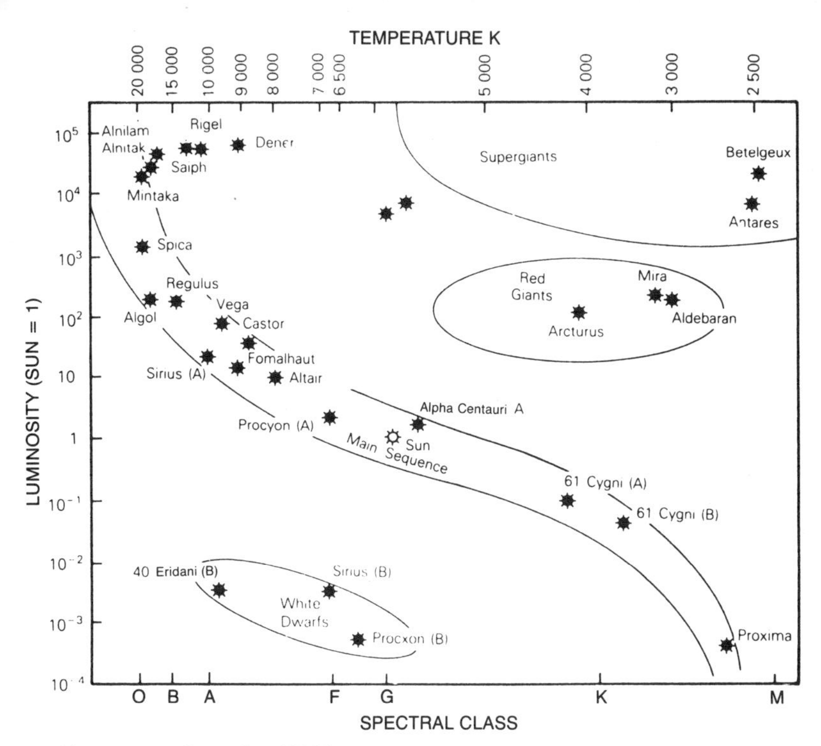

Hertzsprung-Russell or HR Diagram.

The dark lines in the spectra tell us which elements are present; thus hydrogen and helium are dominant in Type B stars, and in Type A it is hydrogen which is most prominent. Metallic lines appear in 'later' types, and in the coolest stars there are lines due to molecules, which could not exist in the hotter stars, because the molecules would be broken up.

When the HR Diagrams were first produced, it was thought that they could give easy clues about a star's evolution. A star would be formed from interstellar material, and would at first be very large, cool and diffuse. Gravity would make it shrink, and it would start to heat up, becoming a red giant. As the shrinking continued, the star would become hot and white, and join the Main Sequence at type B or A. Next it would slide down the Main Sequence, becoming a cool red dwarf before losing all its light and heat.

It all sounded very neat, but it proved to be very wide of the mark. The real problem is that we cannot, in general, see a star evolving, any more than we can see a baby in a pram turn rapidly into a boy. All that astronomers could do was to select certain types of stars and decide whether they were ancient or youthful. Through no fault of their own, they picked wrong. Red giants are not young at all; they are stellar old-age pensioners.

It is true that stars begin their careers by condensing out of the thinly-spread interstellar material, and very often they do so in clusters. What happens next depends chiefly upon the star's initial mass. If it is very low (less than about one-tenth that of the Sun) it will simply shine feebly for a while and then fade away into a dead globe. If it is of greater mass, there will come a time when the core temperature rises to around 10,000,000 °C so that nuclear reactions are triggered off. The star joins the Main Sequence, and settles down to a long period of steady, sober existence. This is the present state of the Sun; it is about 5000 million years old, and is about half-way through its Main Sequence career.

When the supply of available hydrogen 'fuel' is exhausted, the star will have to change its structure. The core will shrink further, and the outer layers will expand and cool; the star will become a red giant, as Betelgeux in Orion is now. Different nuclear reactions will begin, and the star will become unstable. Eventually the outer layers will be blown off, producing what is (misleadingly) termed a planetary nebula, and will gradually dissipate in space. All that will be left of the star will be the core, now made up of the broken parts of atoms packed closely together with little waste of space. The star has become a white dwarf, with a density perhaps 100,000 times that of water. After an immense period, all its light and heat will leave it, and it will become cold and dead — in fact, a black dwarf.

Many white dwarfs are known, the most famous example being the faint companion of Sirius. It would be difficult to detect black dwarfs, for obvious reasons, and it is by no means certain that the universe is yet old enough for any black dwarfs to have formed. We are dealing with time-scales which are incredibly long, even by cosmic standards.

A star of greater mass, over 1.4 times that of the Sun, will suffer a more dramatic fate. When its main nuclear fuel is used up, it will collapse. There will be an implosion, followed by a rebound and an explosion as the star blows most of its material away into space in what is termed a supernova outburst.** For a while the star may shine with a luminosity matching 100,000 million Suns. When the outburst is over, all that remains will be a patch of expanding gas, together with the remnant of the original star — now made up of particles called neutrons — so dense that 1000 million tons of its material could be packed into the bowl of an ordinary pipe.

A neutron star is indeed a bizarre object. It will have a strong magnetic field, and it will spin round quickly — in some cases many times a second, showing that it must be very small; the

** To be precise, this is a Type II supernova. There is another type of supernova outburst, about which I will have more to say in Chapter 13.

diameter may be only a few kilometres. It may also be a source of radio waves. The beams of radiation come out from the magnetic poles, and every time one of these beams sweeps across the Earth we receive a 'pulse' of radio energy. This is why neutron stars are so often referred to as pulsars. The first pulsar was discovered in 1967 by Jocelyn Bell-Burnell, working at Cambridge Observatory in England. The pulses were so rapid, and so regular, that there was a brief period when the researchers thought that they might be picking up artificial transmissions from across the Galaxy.

The best-known supernova remnant is the Crab Nebula, in Taurus (the Bull), which can be glimpsed with good binoculars and is an easy object in a small telescope, though photographs taken with large instruments are needed to bring out its amazingly complex structure. It is the wreck of a supernova which flared up in the year 1054, and was studied by Chinese and Korean stargazers; for some months it was visible in broad daylight. It contains a pulsar, which shows up in giant telescopes as a very faint, flashing object. The Crab is 6000 light-years away, so that the actual outburst took place in prehistoric times.

Supernovae are rare, and the last to be seen in our Galaxy appeared as long ago as 1604. However, in 1987 a supernova flared up in the Large Cloud of Magellan, a mere 169,000 light-years away, and became a conspicuous naked-eye object for a few weeks. It has now faded back to obscurity. At present there is no sign of a pulsar, but probably one does exist, and sooner or later its radio waves will betray it.

Finally, consider a star which is even more massive — say 10 times as massive as the Sun. When the grand collapse starts, nothing can stop it. The star goes on shrinking, and as it does so the escape velocity goes up. Eventually the escape velocity reaches 300,000 km per second — and this is the velocity of light. Not even light can now break free from the doomed star; and if light cannot do so, then certainly nothing else can, because light is the fastest thing in the universe. The star surrounds itself with a sort of 'forbidden zone' into which matter can fall, but nothing — absolutely nothing — can emerge. This is what we call a black hole.

Black holes send out no energy, but they can be tracked down because of their effects upon objects which we can see. One convincing case is that of Cygnus X1, around 6500 light-years away. Here we have a B-type supergiant star, with a mass about 30 times that of the Sun and a diameter of 18 million km, together with an invisible companion which, as we can tell from the orbital movement of the supergiant, is 14 times as massive as the Sun. Presumably the companion is a black hole; it is pulling material away from its companion, and before this material is sucked into the black hole it is so intensely heated that it emits the X-rays which we can detect.

We have no idea what conditions are like inside a black hole, because all the ordinary laws of science break down. It must be said, too, that there are still a few astronomers who question whether black holes can exist at all, and favour some other sort of explanation. However, the evidence is very strong, and it may be that black holes are very common in the Galaxy.

Stellar evolution is highly complicated, and we have to admit that we are still uncertain about many of the finer points, but we are modestly confident that we are on the right track. We have come some way toward an understanding of the life-stories of the stars.

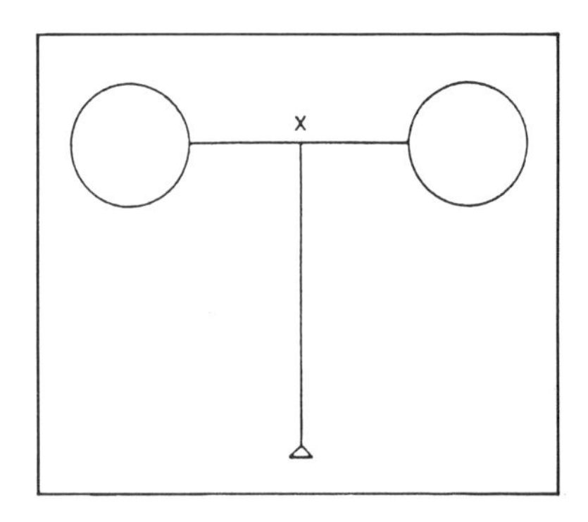

CENTRE OF GRAVITY. If the two bells of a dumbbell are of equal mass, the centre of gravity X will be midway between them. If one mass is greater than the other, the centre of gravity will be displaced toward the more massive bell. This is also the case with the components of a binary system.

Chapter 13
The Variety Of The Stars

When Father Guy Tachard stopped off at the Cape of Good Hope in 1685 and set up a temporary observatory, he made a series of general observations as well as concentrating upon longitude determination. He looked at the Southern Cross and recorded that 'the foot of the Crozier marked in Bayer is a Double Star, that is to say, consisting of two bright stars distinct from each other by about their own Diameter, only much like the northern of the Twins; not to speak of a third much less, which is also to be seen, but further from these two'. The 'foot of the Crozier' is of course Alpha Crucis, the brightest of the stars in the Cross, which Tachard saw to be made up of two — the first double star to be discovered in the southern sky — as well as a third component nearby. The 'northern of the Twins' is Castor in Gemini, already known to be a wide, easy double.

Four years later Alpha Centauri was also found to be made up of two, and today double stars are numbered in tens of thousands. They are of many types. Sometimes the components are perfect twins; sometimes one component is decidedly brighter than the other, as with Alpha Crucis, Alpha Centauri and Mizar in Ursa Major (the Great Bear), which is never well seen from far-southern latitudes but is familiar to every Englishman who takes even a passing interest in astronomy. In many cases one component far outshines the other. For example, the dim companion of Sirius, now known to be a white dwarf, has only 1/10,000 the brightness of its brilliant primary.

Some double stars are mere optical pairs, in which the two components are not genuinely associated, and simply lie near the same line of sight, almost one behind the other, so to speak. Such is Al Giedi or Alpha Capricorni, in the Zodiacal constellation of the Sea-Goat (Chart 18, p. 156). The two stars can be separated with the naked eye; their magnitudes are respectively 3.6 and 4.2. The brighter star is 110 light-years away, but the fainter member lies at 1600 light-years!

Oddly enough, these optical pairs are very much in the minority, and most double stars are physically-associated or binary systems. The components move round their common centre of gravity in periods which range from a few minutes for very close pairs up to many millions of years for pairs which are widely separated, so that all we can really say is that the two components share a common motion in space. The first binary system to be identified was Castor. It was seen to be double by G.D. Cassini in 1678; in 1803 William Herschel announced that the two components were in orbit round their centre of gravity, and we now know that the revolution period is about 420 years. Many catalogues of binaries have been drawn up, and here the Republic Observatory in Johannesburg has been particularly active. Using the 68-cm refractor, major contributions have been made by astronomers such as R.A. Innes, H.E. Wood, W.H. van den Bos and W.S. Finsen. It is a great pity that this work has been suspended — only temporarily, we hope.

With a double, we need to know two facts. First there is the separation, in seconds of arc. There is also the position angle, which is the angular director of the secondary component from the primary, reckoned from 000 at north round by east (090°C), south (180°C) and west (270°C) back to north. With slow-moving pairs, these figures do not alter much. For example the two components of Alpha Crucis are separated by 4.4 seconds of arc at a position angle of 115°, and Father Tachard must have had virtually the same view. On the other hand, the orbital period of Alpha Centauri is only 80 years, so that both the separation and the position angle alter quite quickly. The separation ranges between 22 seconds of arc and only 2 seconds of arc; the real distance between the two is about 5200 million km — 35 times greater than the distance between the Sun and the Earth. The changing separation merely means that we see the system from different angles.

For a double with two equal components, it is probably correct to say that the limiting separation is 2.5 seconds of arc for a 2.5-cm telescope, 1.8 seconds for a 7.6-cm, 0.9 seconds for a 15-cm and 0.4 seconds for a 30-cm. This means that practically any telescope will be capable of separating the components of Alpha Crucis and, almost always, those of Alpha Centauri. Another fine double is Gamma Centauri (Chart 16, p. 150); each component is of magnitude 2.9, and the separation is 1.4 seconds of arc, so that you need an aperture of more than 8 cm to split it well.

Very close binaries cannot be split with any telescope, and here we have recourse to the spectroscope. In some cases the components must be almost touching, and one system, discovered in 1987, has an orbital period of only 11 mins.

Some double stars show lovely contrasting colours. Albireo or Beta Cygni, in the Swan (Chart 21, p. 162) is made up of a golden yellow primary together with a vivid blue companion, while some bright red stars, such as Antares in the Scorpion (Chart 17, p. 155) and Ras Algethi in Hercules (Chart 22, p. 165) have companions which look distinctly green, though admittedly this colour is due largely if not entirely to contrast.

We also have multiple stars. Silhouetted against the Great Nebula in Orion, about which I will have more to say, is the famous Theta Orionis, nicknamed the Trapezium because of the arrangement of its four main components. The classic example of a double-double or quadruple star is Epsilon Lyrae, in the northern constellation of the Lyre (Chart 21, p. 162), close to Vega. It is well to the north of the celestial equator, but can attain a reasonable altitude during evenings in July and August. Keen-eyed people can see that it is made up of two, and a modest telescope will show that each component is again double.

It used to be thought that a binary system was the result of the break-up of a rapidly-spinning single star, but nowadays it seems more likely that the components condensed together out of the interstellar cloud and have always remained gravitationally linked. Rather surprisingly, binaries are more common than single stars such as the Sun.

Most stars shine steadily year after year, century after century, but there are some which brighten and fade over comparatively short periods. Studies of these variable stars are of the utmost importance to astronomers, and each particular type has its own points of special interest.

First there are the eclipsing stars, which are not really variable at all even though they show apparent changes in brightness. A typical example is Zeta Phoenicis, in one of the Southern Birds (Chart 19, p. 158). Normally it shines as a star of magnitude 3.9, but every 1.7 days it fades to magnitude 4.4, remaining dim for only a short while before recovering. Zeta Phoenicis is made up of two components, one brighter than the other; when the brighter member of the pair is partly covered by its companion, we see a fade. Stars of this type are officially termed Algol variables, because the prototype is Algol in Perseus (Chart 7, p. 134), which is much brighter than Zeta Phoenicis but which lies more than 40°N of the celestial equator. With Beta Lyrae (Chart 21, p. 162) the two components are less unequal, and are almost touching, so that variations are always going on. The best southern example is V Puppis (Chart 9, p. 137), which has a range of from magnitude 4.7 to 5.2 and a period of 1.45 days. It lies close to the bright star Regor, or Gamma Velorum.

Beta Doradûs, in the Swordfish (Chart 8, p. 136), is very different. It has a small range, from magnitude 3.7 to 4.1, and its period is absolutely regular at 9.8 days. It is known as a Cepheid, because the prototype star of the class is Delta Cephei, so far north in the sky that it is unobservable from New Zealand and also from most of Australia and South Africa. Many hundreds of Cepheids are known, though not very many are visible with the naked eye.

Cepheids are pulsating yellow supergiants. They are well advanced in their life-stories, so that they have used up their hydrogen 'fuel' and are drawing on their reserves, which makes them unstable. They are of unique importance because the period of variation gives a key to the star's real luminosity. The longer the period, the more powerful the star, so that, for instance, Beta Doradûs (period 9.8 days) is more luminous than Delta Cephei itself (period 5.4 days). Once a star's real brightness is known, its distance can be worked out, so that Cepheids act as invaluable standard candles in space.

Next we have the long-period stars, where the periods amount to many weeks or even a few years. There is no link between period and luminosity, and for that matter neither the periods nor the amplitudes are absolutely constant — in marked contrast with the regular-as-clockwork Cepheids. The prototype long-period star is Mira in Cetus, the Whale. At its best Mira can rise to above the second magnitude, but at minimum it drops to below 10, so that even binoculars will not show it. The mean period is 331 days, so that Mira is visible with the naked eye for only a few weeks in every year. It was, incidentally, the first variable star to be recognized as such — by the Dutch observer Phocylides Holwarda in 1638.

Mira variables are common. Most of them are red, with spectra of type M, N, R or S; amateur observers do sterling work in monitoring them — a task which would be far too time-consuming for the professional. Associated with the Mira stars are the semi-regular variables, also red supergiants, which have smaller amplitudes and less well-marked periods. Betelgeux in Orion is the classic example. The range is from about magnitude 0.2 to 0.9, and there is a very rough period of about 2100 days.

Some variable stars are completely irregular. Such is R Coronae Borealis in the Northern Crown (Chart 20, p. 160), which is usually on the fringe of naked-eye visibility, but which sometimes fades without warning, falling to below magnitude 15 before recovering. What apparently happens is that soot accumulates in the star's atmosphere, and acts as a screen until it is blown away. Other stars remain at minimum for most of the time, but periodically flare by several magnitudes. Such are SS Cygni in the Swan (Chart 21, p. 162) and U Geminorum in the Twins (Chart 6, p. 133), both of which always remain far below naked-eye visibility, and are dwarf binary systems. But the most remarkable of all erratic variables is Eta Carinae, in the Keel of the Ship (Chart 9, p. 137). For a while, during Sir John Herschel's sojourn at the Cape in the 1830s, it shone as the most brilliant star in the sky apart from Sirius, but for more than a century now it has remained below naked-eye visibility.

One evening in 1975 I went outdoors about half an hour after darkness had fallen, to make some routine observations. On the way to my observatory I looked up at the region of Cygnus (Chart 21, p. 162) and saw a bright star where nothing had been visible before. I knew at once what it was: a nova, or 'new star' — a bad name, because a nova is not genuinely new at all, and is merely a formerly dim star which has suffered a tremendous outburst and has flared up to many times its normal brightness.

The star in Cygnus had certainly been absent on the previous night. I telephoned the Royal Greenwich Observatory but, as I expected, I had been forestalled, because the star had been found several hours before by a Japanese observer, Honda. Darkness comes earlier in Japan than in England!

Novae are not too uncommon, and several seen during the past few decades have become quite bright, notably CP Puppis in 1942, which reached magnitude 0.4. Many fainter novae are discovered by amateur astronomers, who know the sky a great deal better than most professionals.

A nova is a binary system, made up of an ordinary star together with a highly-evolved white dwarf. The white dwarf pulls material away from its larger, less dense companion, and material accumulates until the situation becomes unstable. The mass round the white dwarf goes 'over the limit', and there is a mild nuclear outburst, with gas being hurled out at a high speed. At the end of the outburst, the system returns to normal. A few stars, such as the 'Blaze Star' T Coronae Borealis in the Northern Crown (Chart 20, p. 160) have been known to suffer more than one outburst; T Coronae itself flared up to the second magnitude in 1866 and again in 1946. The only other 'recurrent nova' to attain naked-eye visibility is RS Ophiuchi, in the Serpent-bearer (Chart 22, p. 165), which suffered outbursts in 1901, 1933, 1958 and 1967.

Finally we have supernovae, which are of two classes. In a Type I supernova, we are again dealing with a binary system in which one component is a white dwarf, made up largely of carbon. As with ordinary novae, the white dwarf collects material from its companion, but this time the mass goes on accumulating until it exceeds 1.4 times the mass of the Sun. There is then an incredibly violent explosion. The white dwarf literally blows itself to pieces, and is completely destroyed. For a while the luminosity may be a 100,000 million times that of the Sun, equal to the combined output of all the stars in a typical galaxy!

I have already referred to Type II supernovae, which are quite different. A Type II is due to the collapse of a very massive star, so that the end product is a neutron star plus a cloud of expanding gas which gradually drifts away into space. The 1987 outburst in the Large Cloud of Magellan was a Type II; we await the detection of a pulsar there. Astronomers would dearly like to study a supernova in our own Galaxy (provided that it is not too close!) and perhaps the best potential candidate is Eta Carinae, which is very massive and unstable and has a limited life-span. It may blow up at any time — perhaps tomorrow, perhaps not for a million years, but certainly within two million years. The nearest red supergiant is Betelgeux, at 310 light-years. No doubt it will 'go supernova' eventually, and if so it will shine on Earth as brilliantly as the full moon.

Certainly there is plenty of variety among the stars. They are fascinating companions; all have their own quirks and peculiarities, and they can never be dull.

Chapter 14
Star-Clusters And Nebulae

Our year begins on the first of January, a date which is not marked by any particular event in the sky. Before white men arrived in New Zealand, the Maori started their year with the first appearance of the Pleiades in the dawn sky, and certainly this lovely cluster makes a fitting marker.

It lies in the constellation of Taurus, the Bull (Chart 4, p. 130). At first glance it looks like a misty patch, but closer inspection shows that it is made up of stars. The nickname of the 'Seven Sisters' is appropriate. Years ago I carried out an experiment, asking television viewers to examine the cluster and note how many stars they could make out on a clear night; the average was indeed seven, though some people could see more (I have one colleague who can glimpse as many as 16 under excellent conditions). Binoculars show that there are dozens of stars in the group, and the total number of Pleiads is between 400 and 500. Most of the leaders are hot and bluish-white. The brightest member of the cluster, Alcyone or Eta Tauri, is about 350 times as luminous as the Sun.

In 1781 a French astronomer, Charles Messier, drew up a catalogue of clusters and nebulae, and gave them numbers; the Pleiades is therefore M.45. (Actually, Messier was not in the least interested in nebulous objects. He was a comet-hunter, and compiled his list merely to avoid confusing clusters and nebulae with comets.) His catalogue includes various loose or open clusters, though there were also some notable omissions. For instance, Messier ignored the Hyades, which extend in a sort of V-formation from the bright orange-red star Aldebaran, also in Taurus, presumably because there was absolutely no danger that the Hyades could be mistaken for a comet.

Other bright open clusters are M.44, Praesepe, in Cancer (Chart 11, p. 143) and M.6 and M.7 in Scorpius (Chart 17, p. 155), all of which are very easy to see with the naked eye, but for sheer beauty it is difficult to match the Jewel Box, round Kappa Crucis in the Southern Cross (Chart 1, p. 120). It lies next to Beta Crucis, and is in the same binocular field. Most of its stars are hot and bluish-white, but there is one red supergiant which stands out at once. The Jewel Box has no Messier number, because it is much too far south to be seen from France, where Messier spent his whole life, but it is included in the New General Catalogue, drawn up by the Danish astronomer J.L.E. Dreyer; this was in the 1880s, so that by now the 'new' catalogue is over a century old. The Jewel Box is listed as NGC 4755. It is about 7700 light-years away, with a central region 25 light-years across.

The stars in an open cluster must have been formed from the same cloud of material at around the same time. They will not keep their identities for ever, because as time passes by they will be disturbed by nearby stars and will gradually disperse. This is not true of systems of entirely different type — the globular clusters, of which the prime example is Omega Centauri (Chart 16, p. 150). It is easy to find, because it lies in line with Beta Centauri, the second of the two Pointers, and the second-magnitude star Epsilon Centauri. With the naked eye it looks like a misty patch of about the fourth magnitude, but binoculars will show that the outer parts are starry, and an adequate telescope can resolve it beautifully. It contains at least a million stars, and is prominent even though it is around 17,000 light-years away. It was recorded by Ptolemy, last of the great astronomers of Classical times, and must also have been seen by the early Maori and Aborigines, though they did not regard it as nearly so important as the Pleiades. Only two other globulars are distinctly visible without optical aid; 47 Tucanae, almost silhouetted against the Small Cloud of Magellan (Chart 19, p. 158) and M.13 in Hercules, in the north of the sky (Chart 22, p. 165).

More than a hundred globular clusters are known in our Galaxy, forming a sort of 'outer framework' to the main system. Most of them contain short-period variable stars, and it was by using these as standard candles that Harlow Shapley, during World War I, was able to measure their distances and give the first reasonably accurate estimate of the size and shape of the Galaxy itself. As well as Cepheids, there are the so-called RR Lyrae stars, which also pulsate regularly and have very short periods. They are just as helpful as the Cepheids, because all of them seem to be about 100 times as luminous as the Sun.

Near the core of a globular cluster the stars are crowded together much more closely than in our own part of the Galaxy. If we lived on a planet moving round a star in the middle of, say, Omega Centauri, the sky would be ablaze. There would be many stars brilliant enough to cast shadows, and many of them would be red, because globular clusters seem to be so ancient that their leading stars have used up their main 'fuel' and have evolved off the Main Sequence to become red giants and supergiants. But our globular-cluster astronomer would probably know very little about the outer universe!

Two of Messier's objects, Numbers 57 and 76, are quite different. They are known as planetary nebulae, which is a misnomer inasmuch as a planetary nebula is not truly a nebula and has absolutely nothing to do with a planet. It is an old star which has thrown off its outer layers, so that it is surrounded by a shell of tenuous gas. When we look at it, we see a greater thickness of gas at the edge of the shell, so that the object looks like a ring with a central star. The brightest planetary nebula is NGC 7293, in Aquarius, the Water-bearer (Chart 18, p. 156); for some reason or other Messier overlooked it. It is nicknamed the Helix, and is actually brighter than M.57, the Ring Nebula in Lyra (Chart 21, p. 162), which is too far north to be well seen from southern countries. There are plenty of planetary nebulae, though most of them are faint. The central stars are small, hot and very dense, because they are well on their way to becoming white dwarfs.

Note, too, No.1 in Messier's list. This is the Crab Nebula, the only supernova remnant which he recorded. It was discovered by John Bevis in 1731, but not until much later was it definitely identified with the star of 1054.

M.42 is the Orion Nebula, in the Hunter's Sword, just above the three stars of the Belt. This is a huge gas-cloud, 30 light-years across and well over 1000 light-years away. It is easy to see with the naked eye, and a small telescope will show the four main components of the Trapezium, the multiple star Theta Orionis. These stars really are associated with the nebula; they lie near the Earth-turned edge of it, and they are very hot, so that they not only illuminate the gas and dust but also make the nebular material send out a certain amount of light on its own account. The main constituent is, predictably, hydrogen, together with a great deal of 'dust'. Yet the density is almost incredibly low; David Allen has pointed out that if you could take an inch core sample right through it, the total amount of material collected would weigh no more than a dollar coin.

And yet fresh stars are being born inside the Orion Nebula, which qualifies as a true stellar nursery. At first the fledgling stars are unstable, and vary irregularly; they are going through what is called the T Tauri stage, and have yet to settle down on the Main Sequence. There are other objects inside the nebula, too. One of these is called BN, short for the Becklin-Neugebauer object. We cannot see it, and we never will, because its visible light is blocked by the nebular dust, but we can pick up its infra-red emissions, and we know exactly where it is. It is now thought to be an immensely powerful star, destined to remain for ever cut off from the rest of the visible universe. It will not live for long enough to 'burn a hole' in the nebula and send its light outward. M.42 is only the brightest part of a huge molecular cloud covering most of Orion, but it is striking because it is being lit up by the Trapezium stars.

If there were no suitable stars nearby, the nebular material would not shine; and dark nebulae do indeed occur. The best-known example is the Coal Sack in the Southern Cross, which blots out the light of stars beyond. It lies close to Alpha and Beta Crucis, and is well seen in binoculars. For all we know, the far side may be illuminated, so that from a different vantage point in the Galaxy the Coal Sack may well look bright.

Open clusters, globulars, gaseous nebulae, planetaries, supernova remnants — all these are contained in our Galaxy. There is also a tremendous amount of thinly-spread interstellar matter, mainly hydrogen together with 'dust'. It has been found that the clouds of cold hydrogen emit radio waves at a certain definite wavelength — 21.1 cm — and by plotting the distribution of these clouds it has been possible to confirm that the Galaxy is spiral in form, like a huge Catherine wheel. This is no surprise, because far away in space we can see many other spiral star-systems.

We cannot see through to the centre of the Galaxy because there is too much obscuring matter in the way; it lies beyond the glorious star-clouds in Sagittarius (Chart 17, p. 155) which are so much in evidence during evenings in winter. Radio waves can slice through the star-clouds, and there is certainly a strong radio source in the very heart of the Galaxy. It may be a massive black hole, though as yet we cannot be sure.

Up to less than eighty years ago it was still thought that our Galaxy was unique, and that it made up the entire universe. We know better today, and we know that just as the Earth is an unimportant planet and the Sun is an unimportant star, so the Milky Way system is an unimportant galaxy. The more we learn, the less significant we seem to become!

Chapter 15
The Flight Of The Galaxies

Two of the most remarkable objects in the sky are the Magellanic Clouds. They are named after the great Portuguese explorer, but must have been known since very early times, because they are prominent features with the naked eye. Even moonlight will not drown the Large Cloud, which lies in the region of Dorado, the Swordfish (Chart 8, p. 136). The Small Cloud, most of which lies in Mensa, the Table (Chart 1, p. 120) is also very evident. Both are independent galaxies. The Large Cloud is 169,000 light-years away, while the Small Cloud is slightly more remote. It is now thought to be a double system, with one component lying almost behind the other.

Astronomically the Clouds are of unique importance, and it is partly on their account that most new observatories are being set up in the southern hemisphere rather than in Europe or the United States, where the Clouds never rise. They are close to the south celestial pole, and over New Zealand and most of Australia and Southern Africa they remain permanently above the horizon.

The Clouds contain variable stars and it was by studying the short-period variables in the Small Cloud, in 1912, that Henrietta Leavitt discovered the link between the periods and luminosities of these stars. The variables with the longer periods looked the brighter, and since all were essentially at the same distance from us it followed that they really did have higher luminosities. (Of course, the distances are not identical, but the differences are negligible in relation to the remoteness of the Cloud itself.) Novae flare up from time to time, and there was of course the 1987 supernova in the Large Cloud, which remained visible with the naked eye for some weeks and altered the superficial appearance of the whole of that part of the sky.

The only other galaxy clearly visible with the naked eye is M.31, the Andromeda Spiral, whose high northern declination makes it awkward to observe, though it does actually rise above the horizon even from Invercargill. It is a true spiral, though it lies at a narrow angle to us and its full beauty is lost. However, by no means all galaxies are spiral in form. Some are elliptical, some spherical and some irregular. There are also 'barred spirals', in which the arms issue from the ends of a sort of bar passing through the centre of the system. The sizes also show a wide range. Some galaxies are much larger than our own, while others are so small that they are not much more populous than globular clusters. Edwin Hubble produced the first system of classification, and what is nicknamed his 'tuning-fork' diagram is still valid.

The status of the so-called 'starry nebulae' remained uncertain right up to the 1920s, when Hubble carried out some vitally important work with the Mount Wilson 100-in (254-cm) reflector, which was then not only the largest telescope in the world but was in a class of its own. Hubble traced Cepheid variables in M.31 and other spirals, and measured their distances simply by observing their behaviour and applying Henrietta Leavitt's period-luminosity law. As soon as he did this, he realised that the systems are so far away that they cannot possibly belong to the Milky Way. He gave the distance of M.31 as 900,000 light-years, later reduced to 750,000 light-years. This was an under-estimate, because Hubble did not know — and could not know — that there are two types of Cepheids, one twice as powerful as the other, and quite understandably he had picked wrong. We now know that the Andromeda Spiral is 2.2 million light-years away; apart from the Magellanic Clouds and a few dwarf systems, it is the nearest of all the galaxies. With galaxies more than a few tens of millions of light-years away, the Cepheids fade into the general background, and other methods of distance-gauging have to be used, but we are modestly confident that our results are not greatly in error.

Galaxies tend to congregate in groups or clusters. Our own Galaxy is a member of the Local Group, which includes the Clouds, the Andromeda Spiral, the small spiral in Triangulum, possibly another large system which is heavily obscured by dust lying in the main plane of the Milky Way, and more than two dozen assorted dwarfs. In fact the closest of all the galaxies, a dwarf spheriodal system in Sagittarius, was discovered in 1994 by Roderiga Ibata, working with a group at Cambridge University in England. It is 80,000 light-years from us, but only 50,000 light-years from the centre of the Galaxy, because it lies on the far side. It has not previously been recognized because it lies in a rich area of the sky. It ranks as a satellite of our Galaxy, and is being torn apart and now extends over at least 10,000 light-years; over the next few hundred million years it will slowly disperse and lose its separate identity. To quote Dr. Michael Irwin, one of the Cambridge team, 'It is being gobbled up by our own Galaxy.'

Other groups lie beyond. For example the Virgo Cluster, at around 50 million light-years, contains over 1000 systems, notably the giant elliptical galaxy M.87 in Virgo (Chart 15, p.149), which is moreover a strong radio source and may well hide a gigantic black hole in its centre.

Hubble also made another discovery which is of paramount importance in modern astronomy. All the clusters of galaxies are moving away from each other — and the further away they are, the faster they are going, so that the entire universe is in a state of expansion. Proof of this comes from studies of their spectra.

Next time a car passes you with the driver's hand clamped firmly on the horn, listen to the note. When the car is approaching, the sound will be high-pitched. After it has passed you and begun to recede, the pitch will drop. There is no mystery about this behaviour, because sound is a wave motion. When the car is approaching, more sound-waves per second reach your ear than if the car were standing still, so that to all intents and purposes the wavelength is shortened. When the car is receding, fewer sound-waves per second enter your ear and the result is an apparent increase in wavelength.

Light, too, is a wave motion and the same principle applies. With an approaching object the wavelength is shortened, and the light-source appears 'too blue'. With a receding object the wavelength is increased and the source appears 'too red'. This is the Doppler effect, named in honour of the Austrian physicist who first drew attention to it as long ago as 1842.

The actual colour-change is very slight and for everyday speeds it is absolutely inappreciable — do not expect a car's headlights to change from blue to red as they flash by! But with a star, the effect shows up in the spectrum. If the star is approaching, all the dark lines crossing the rainbow band are moved over toward the blue or short-wave end. With a receding star, the shift is to the red. The greater the shift, the greater the relative velocity between the star and ourselves.

Using this method, astronomers were able to measure the 'radial' (i.e. toward-or-away) motions of many stars. Since a galaxy is made up of a vast number of stars, its spectrum is bound to be something of a jumble, but the main dark lines can be made out easily enough and Hubble found that all the galaxies beyond our Local Group showed red shifts. Each galaxy was receding from us, and the speed of recession became greater with increasing distance. This also gave a means of estimating the distances of galaxies; all that had to be done was to measure the extent of their red shifts.

It is not true to say that our Galaxy is particularly unpopular, and we are in no special position in the universe. Each group of galaxies is flying away from each other group, so that the only galaxies which are not receding from us are those of the Local Group.

Modern techniques allow us to examine systems which are

so far away that the light takes over 12,000 million years to reach us. The most remote objects so far found are quasars, which were first tracked down in 1963.

Radio astronomy provided the clue. One radio source, known as 3C-273 (because it was the 273rd entry in the third Cambridge catalogue of radio sources) was surprisingly strong, but for a long time it could not be identified with any optical object because its position was too uncertain. This is where the Parkes radio telescope came in. On 5 August 1962 the radio source was hidden or occulted by the Moon, and the event was monitored from Parkes. Since the position of the Moon was known with very high accuracy, the position of 3C-273 could be found from the moment its radio signals were cut off. From this, the source could be identified with what seemed to be a faint blue star of below the 12th magnitude.

The Parkes team sent their results to Palomar in California, where Maarten Schmidt used the 200-in (5-m) Hale reflector to obtain an optical spectrum. The result was amazing — 3C-273 was not a star at all. Its spectrum was quite different and the dark lines showed tremendous red shifts, proving that the object was very remote. It was classed as a Quasi-Stellar Radio Source, a term subsequently shortened to quasar, though it has now been established that many quasars are not radio emitters so that the term QSO or Quasi-Stellar Object is coming into favour.

Quasars are very small compared with galaxies. Hundreds of them are now known and they are immensely luminous. All lie far beyond our Local Group and even the nearest is several thousands of millions of light-years away. It seems that a quasar is the core of a very active galaxy, probably powered by a vast, massive black hole.

Quite recently, grave doubts have been expressed about all our distance-measurements beyond our own immediate part of the universe. We depend mainly upon the red shifts in the spectra, but suppose that these are not, after all, pure Doppler effects? This is what is believed by several very eminent astronomers, including Dr. Halton Arp of the United States (now working in Germany), Sir Fred Hoyle and Dr. Geoffrey Burbidge. Arp has produced pictures of galaxies and quasars which are clearly aligned, but which have totally different red shifts. If his ideas are correct, then many of our current theories will have to be drastically revised. This is highly unpopular with most astronomers, and indeed Arp was forbidden to continue using the Mount Wilson reflector because the results he was obtaining were too embarrassing. We must wait and see . . . In any case, there are several fundamental questions which we cannot yet answer. In particular, how large is the universe? Secondly, how did it begin?*

Either the universe is finite, or else it is not. If it is finite, then what lies outside it? To say 'Nothing' is merely evading the issue. But if the universe is infinite, we have to visualize something which spreads out for ever, and our brains are unequal to the challenge. However, at least we may be able to define the limits of the observable universe. As we have noted, the recessional velocity of a galaxy (or a quasar) increases with distance. If this law holds good, we must eventually come to a point at which the object is racing away at the full speed of light. We will then be unable to see it, and we will have reached the edge of the observable universe, though not necessarily of the universe itself.

It is believed that this limit lies somewhere between 15,000 million and 20,000 million light-years, with a marked preference for the lower value. And from this, it seems logical to assume that the universe in its present form is around 15,000 million years old.

According to current theory, the universe — everything; space, time, matter, radiation — came into existence at one set moment, in a 'Big Bang'. To ask 'where' this happened is meaningless, because if space were created at the same instant, the Big Bang happened 'everywhere'. At once the fledgling universe began to expand and the initially high temperature cooled. Atoms and then molecules were formed; galaxies condensed, then stars, planets, and finally you and me. If we accept the concept of the Big Bang, we can work out a complete evolutionary sequence. What we cannot do is to explain the Big Bang itself, and here we have to admit that we are completely at a loss.

Support for the Big Bang idea came in 1965 when two American radio astronomers, A. Penzias and R. Wilson, traced down very weak radiation coming from space from all directions all the time. This indicates a general temperature of about 3° above absolute zero (absolute zero being the coldest temperature there can possibly be) and is the last remnant of the Big Bang. There have been further studies since, largely with an artificial satellite called COBE (the Cosmic Background Explorer) and the picture does seem very convincing, though there are still some astronomers who claim that the Big Bang never happened and that the universe has always existed.

What of the future?

If the average density of matter in the universe is greater than a certain critical value (about one atom of hydrogen per cubic metre) then the galaxies will gradually slow down, stop and then start to come together again, so that in perhaps 80,000

* For the rest of this chapter I propose to follow the official theory, not Halton Arp's, even though I have an inner feeling that Arp is right!

million years there will be a new Big Bang and the cycle may be repeated. But if the density is lower than this critical value, the expansion will never stop, and the groups of galaxies will go on receding from each other until all contact between them is lost. On the first picture, the universe resembles a clock which is being continuously re-wound. On the second, the clock will simply be allowed to run down and stop.

The amount of matter which we can actually see is quite inadequate. Put all the galaxies, stars and interstellar material together and there is not nearly enough mass to halt the expansion. But by studying the ways in which the galaxies behave, there is every reason to believe that most of the mass of the universe is made up of material which we cannot see at all. The nature of this 'dark matter' is a complete mystery at the moment, but there seems little doubt that it exists, and it may well be enough to draw the galaxies back.

This is really as much as we can say as yet. The new telescopes will enable us to probe still further into the depths of space, and they may give us at least some of the answers, but we still have a long way to go before we can claim to have even a reasonable understanding of the universe in which we all live.

Chapter 16
Life In The Universe

In 1991 the International Astronomical Union held a General Assembly at Buenos Aires, in Argentina.* One meeting was concerned with what is called SETI, the Search for Extra-Terrestrial Intelligence, and there were even guide-lines laid down to say what should be done if any sign of life were detected. Anything of the sort would have seemed like pure science fiction only a few decades ago.

Certainly it seems conceited to suggest that *Homo sapiens* is unique. Our Sun is one of 100,000 million stars in our Galaxy. We know that there are thousands of millions of other galaxies, and in all this host it does not seem reasonable to believe that only our Sun is attended by a peopled planet. But saying this is one thing, and proving it is quite another. First, we have to establish that planets of other stars really exist.

This is by no means easy. A planet is very tiny compared with a normal star, and has no light of its own, so that it will be excessively faint and hard to detect. We can now hope for much from the Hubble Space Telescope. As almost everyone knows, it was launched with a faulty mirror which caused the alteration of some observational programmes and the cancellation of others. In December 1993 a superb 'repair mission' was carried out by a team of astronauts, with the result that the telescope is now every bit as good as had originally been hoped. It may even be able to detect large planets associated with nearby stars. On terra firma, there are twin Keck telescopes on Mauna Kea in Hawaii; one of the pair is complete — it has a 9.8-m mirror, made up of 36 segments fitted together to make the correct optical curve — and when its twin is ready, in 1995, the pair should be able to detect the headlights of a car separately from 25,000 km. But there are also other ways of detecting extra-solar planets.

A massive planet orbiting a lightweight star would pull on the star and make it 'wobble' very slightly. Effects of this sort have been detected for several stars within 12 light-years of us,

* General Assemblies are held every three years in different countries. So far there has been one Assembly in Australia, but none in New Zealand or South Africa. In Buenos Aires I was elected honorary editor of the IAU Newspaper. I am glad to say that I managed to produce it on time every morning, even when the Conference Centre burned down!

but the perturbations are so slight that they are bound to be uncertain. In any case, all we could hope to do by this so-called astrometric method would be to track down planets which are much more massive than the Earth.

There was an interesting development in 1983 with the launch of IRAS, the Infra-Red Astronomical Satellite, whose task was to map the entire sky at infra-red wavelengths. Though IRAS operated for less than a year, it was a tremendous success. Early in its career, while the instruments on board were still being calibrated, it was found that the brilliant star Vega is associated with cool material which might well be planet-forming. The same was true of various other stars, notably Fomalhaut in the Southern Fish (Chart 18, p. 156) which is almost overhead during evening in spring.

A particularly significant case was that of Beta Pictoris in the little constellation of the Painter (Chart 19, p. 137). It shines as a star of magnitude 3.8; it is white with an A-type spectrum; it is 78 light-years away, and 68 times as luminous as the Sun. Previously it had been regarded as an ordinary star, and had not even been given an individual name, but IRAS found that there was something very special about it. Like Vega and Fomalhaut, it had a huge 'infra-red excess'.

When Beta Pictoris was examined with the 1.5-metre Irénée du Pont telescope at the Las Campanas Observatory in Chile, it was found that the cool material could actually be recorded. Apparently there is a disk of material extending out to almost 80,000 million km from the star. We see it practically edgewise-on, and it may be no more than a few hundred million years old. There may be 'depleted regions' in it, suggesting that parts of the cloud have been swept up by orbiting planets. Then, in March 1993, astronomers using a 30-metre radio telescope in France established that there is a similar disk of material associated with Fomalhaut. This time the diameter of the disk is more than 400 times the distance between the Earth and the Sun.

This is not to say that Beta Pictoris and Fomalhaut are planetary centres, but it is not unlikely, in which case it is logical to assume that Solar Systems are fairly common in the universe.

Next, what are the chances of our being able to establish contact with extra-terrestrial beings? Presumably life, wherever it exists, must be made up of the same elements as we are, because we know that the make-up of the universe is the same everywhere; the elements in a remote galaxy or quasar are the same as those in your bathwater. Of all types of atoms, only carbon can combine in a way which leads to the formation of the molecules needed for life, in which case all life must be carbon-based and we can forget all about BEMs or Bug-Eyed Monsters made of pure gold and living happily on a scorching hot world with a sulphur dioxide atmosphere. This is not to suggest that an extra-terrestrial must look humanoid. He or she, or it, may well have several legs and arms, for instance, but the basic material will be the same. (After all, there is not much outward similarity between a man and an earwig, though both are carbon-based.)

For the moment we can disregard all ideas about physical travel beyond the Solar System. Even if it were possible to go at the speed of light (which, according to relativity theory, it is not), a journey to the nearest stars which are at all like our Sun would take years. We can also ignore the lurid stories about flying saucers flown by little green men. It is not impossible that we might be visited by aliens from a far-away civilization much more advanced than ours, but there is not a shred of real evidence that it has happened yet.

Therefore, the only plausible method is to use radio, since radio waves travel at the same speed as light. We might be able to pick up signals which are sufficiently rhythmical to be interpreted as artificial, and in fact the first serious attempt was made as long ago as 1960, when Frank Drake and his colleagues used the radio telescope at Green Bank, in West Virginia, to carry out a survey. They concentrated upon a wavelength of 21.1 cm, because this is the wavelength of the radiation emitted by the clouds of cold hydrogen spread through the Galaxy, and it was reasonable to think that radio astronomers, wherever they might be, would know about it. The main targets were two stars, Tau Ceti and Epsilon Eridani, both of which are visible with the naked eye (Charts 24, p.169 and 8, p.136 respectively). Both are smaller and cooler than the Sun, but are of the same general type, so they could well have planetary families. Both are between ten and 12 light-years away.

Drake's foray — nicknamed Ozma, after the fake wizard in Frank Baum's famous novel — led to no results, and neither did later attempts made elsewhere. Since 1992, however, there has been renewed interest and an elaborate search programme has been started using, among other instruments, the radio telescope at Tidbinbilla in Australia. The chances of success are not very high, and in any case the results are bound to be limited. If you send out a message to (say) Tau Ceti in 1993, it will arrive there in 2005, so that if some obliging Tau Cetian operator replies promptly the reply might be expected in 2017. This means a total delay of 24 years, making quick-fire repartee rather difficult. Moreover, it is rather too optimistic to expect 'aliens' to speak English, and the only way to convey any sensible message is by some sort of mathematical code. But already we are verging on science fiction and we must await future developments.

There have been periodical suggestions that we should avoid sending rocket probes beyond the Solar System in case they attract attention from alien races who would then consider taking us over. I see no justification for this. It will take one of our rockets many centuries to go anywhere near another possible planetary system, and moreover radio astronomers there, if they exist, will already know about us, because they will have picked up our broadcasts. In any case, any civilization advanced enough to master interstellar travel will have long since abandoned all ideas of warfare — otherwise they would have wiped each other out, as we on Earth are in danger of doing today. Personally I would welcome the arrival of interstellar visitors, because I am confident that they could teach us a great deal!

Meanwhile our direct exploration of the universe is limited to the Solar System, and will remain so unless we make some fundamental 'breakthrough' about which we cannot even speculate as yet. This may happen tomorrow, in a year's time, in a hundred years, a thousand years, a million years — or never, I wish I knew.

Part 2
The Sky Throughout The Year

Chapter 17
Stars Of All Seasons

Sir John Herschel, the great astronomer who took a large telescope to the Cape of Good Hope over a century and a half ago in order to make the first really systematic study of the far-southern stars, once said that the constellation patterns had apparently been drawn up 'so as to cause as much confusion and inconvenience as possible'. One can sympathize with his point of view. Ptolemy, the last of the great Greek astronomers, recognized 48 constellations, most of which are fairly distinctive, but some of the more recent additions to the sky-map are not. Moreover, there was a time when almost every astronomer seemed intent on making his own personal contribution, and some catalogues included obscure groups with barbarous names such as Officina Typographica, Honores Fredericii and Sceptrum Brandenburgicum! Mercifully, these have now been erased; even so, some of the accepted groups seem unworthy of separate names. This is particularly true of the relatively barren area round the south celestial pole.

The system I have always adopted for star recognition is what I call the signpost method. It involves starting off with a very few groups which can hardly be mistaken, and using these as pointers to locate all the rest. There are three particularly useful signposts: Crux Australis (the Southern Cross) in the south, Orion at the equator, and Ursa Major (the Great Bear) in the north. I hope that the method will be found acceptable, so let me give you just one example. Assume that we have found the Southern Cross, and want to identify the brilliant white star Achernar in Eridanus (the River), which is on the far side of the pole. Follow along the line of the 'longer axis' of the Cross (as shown in the diagram) and continue it until you come to the first really bright star; this must be Achernar. If the line reaches the horizon beforehand, as will happen at times, you will know either that Achernar is genuinely out of view, or else that it is hidden by low-lying haze or cloud.

In fact, things are slightly easier in northern latitudes than

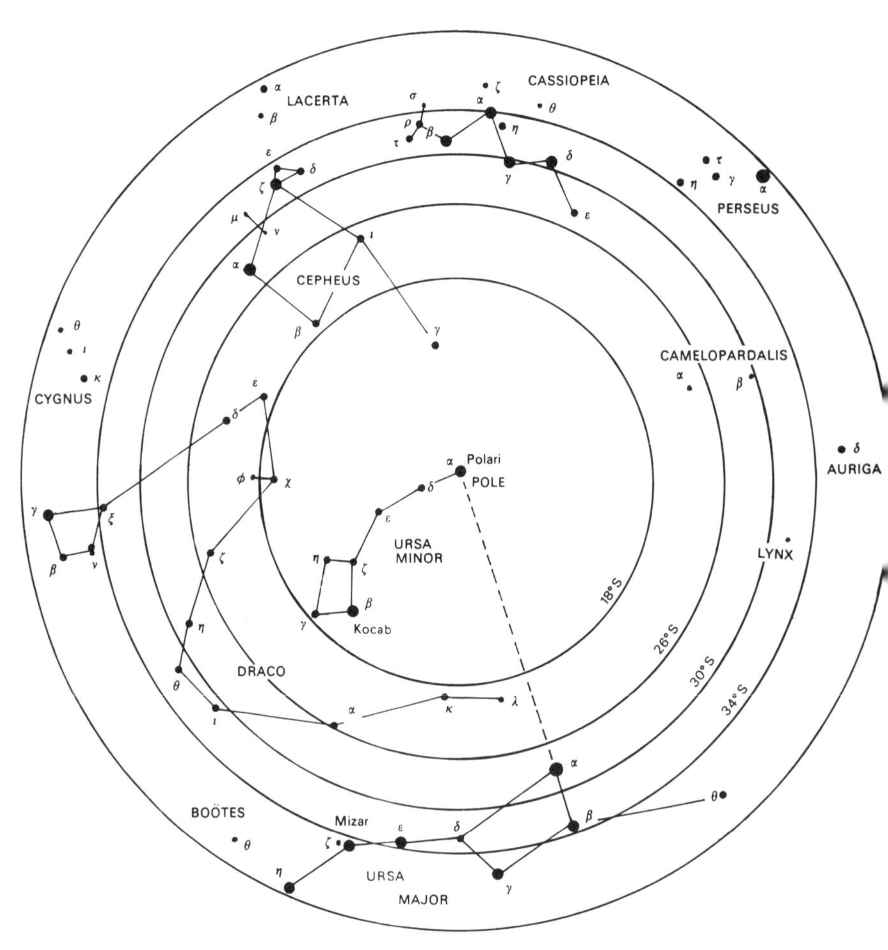

The far-northern stars, with various 'visibility limits'.

in the south, because we can make use of Ursa Major, which is both large and distinctive. Its seven chief stars make up the Plough or Big Dipper, familiar to every Briton. However, it is always very low from most of Australia and South Africa, and from New Zealand it cannot be seen at all. The same is true of the other really prominent far-northern constellation, Cassiopeia, whose five principal stars make up a distorted W or M form. For the sake of completeness I have given a chart of the north polar region, with the limits of visibility for various latitudes ranging from 46°S (Invercargill) through to 18°S (Harare, in Zimbabwe). From this you can see that Dubhe in Ursa Major just rises from Johannesburg (latitude 26°S) but not from Auckland (37°S).

Actually, the north polar area does not contain very much of interest, and the only groups inaccessible from southern

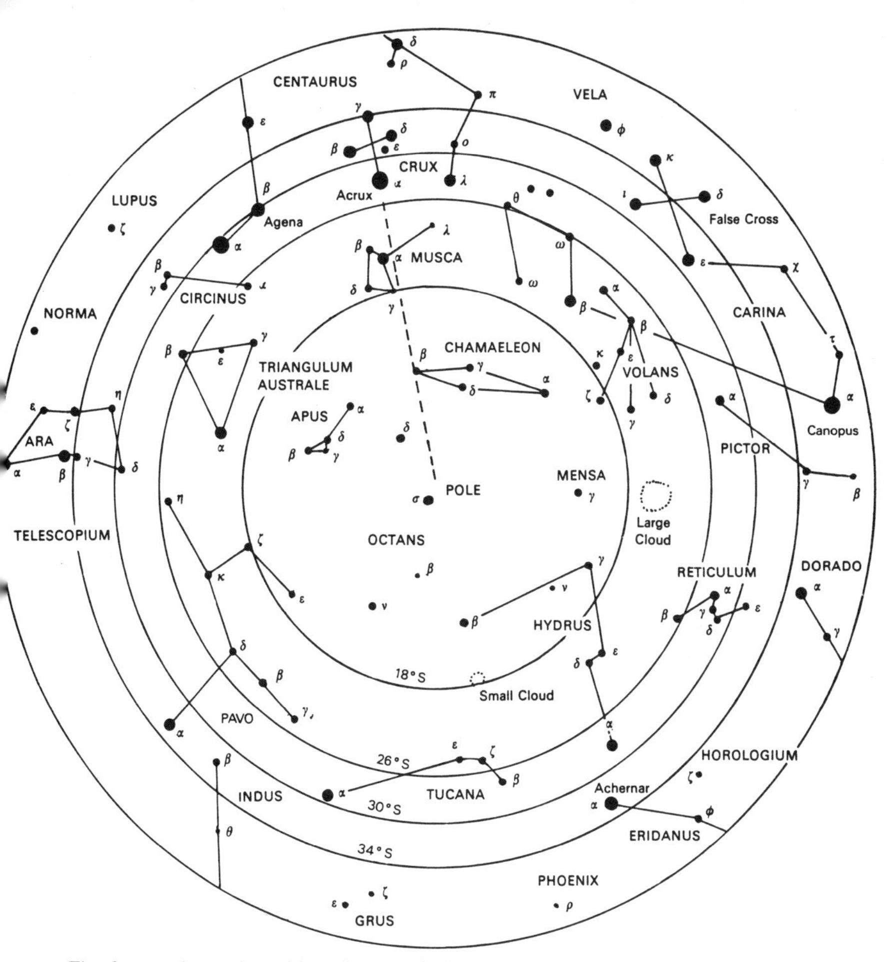

The far-southern sky, with various 'visibility limits'.

countries are Ursa Major, Cassiopeia, and Ursa Minor (the Little Bear) which contains Polaris, the north pole star. There are few important telescopic objects, and the loss to the southern observer is not unbearable, though for Britons the situation is different; the far southern stars which remain permanently out of view are among the most interesting and spectacular in the entire sky!

The south polar chart is shown in the next diagram. Again I have put in dashed lines, but this time to give the limits of the circumpolar zone. For example, from Cape Town and Adelaide (latitude 34 to 35°S) Canopus drops briefly below the horizon during its 24-hour circuit of the pole, whereas Alpha Centauri does not, and Gamma Crucis, the red star of the Cross, is a borderline case. All these are circumpolar from New Zealand, though from Auckland it is true that Canopus at its lowest is so near the horizon that it is not easy to see.

The charts which follow can be used for virtually the whole area, bearing in mind that there are some minor adjustments to be made because there is a difference of 20° in latitude between Invercargill and Johannesburg, while Ursa Major rises well clear of the horizon from Darwin but can never be seen at all from Dunedin. There is an easy way to work out the limits of visibility, so let me give a couple of examples which involve nothing more than simple subtraction.

The co-latitude of any point on the Earth's surface is given by subtracting its latitude from 90°. Thus the co-latitude of Wellington (latitude 41°S) is 49, because 90-41 = 49. This means that any star north of declination 49°N will never rise from Wellington, while any star south of declination 49°S will never set. Consider Canopus, declination 53°S. This is south of the magic figure of 49, so that Canopus will be circumpolar. The brilliant blue star Vega lies at declination 39°N; this also is below the limit, so that from Wellington it will rise, though it will never attain an altitude of more than 10°. (I am not taking account of what is termed refraction, which is the bending of light-waves by the Earth's atmosphere; for a star close to the horizon the effect is quite appreciable. However, I am not aiming to do more than give general principles.)

In the chart I have given a list of the first-magnitude stars, with their declinations, and also a few other stars of exceptional interest. I have also given the latitudes and co-latitudes of some major centres in New Zealand, Australia, South Africa and South America, in all cases to the nearest degree. Armed with these figures, it is easy to work out whether a star is available or whether it is circumpolar. For instance:

You are in Auckland; co-latitude 53°. What will be your views of Capella (dec. 46°N), Alpha Crucis (63°S) and Dubhe (62°N)? Answers: Capella will rise, Alpha Crucis will be circumpolar, Dubhe will never be seen.

You are in Brisbane, latitude 27°S, co-latitude 63°S. From here Capella will rise; Alpha Crucis will touch the horizon when at its lowest; Dubhe will just bob into view, but you will be very lucky to see it!

Obviously it makes sense to start our tour of the sky with Crux Australis (the Southern Cross), because it is circumpolar from New Zealand and is generally on view from most of Australia and South Africa, whereas of our two other main 'skyposts' Orion is out of view altogether during much of the winter and Ursa Major is never sufficiently high up to be of much use as a direction-finder. It is rather surprising to find that Crux is actually the smallest constellation in the sky; it is not an 'ancient' group, since it was not formed until 1679 by an

Stars of the First Magnitude

Star		*Magnitude*	*Dec°*	*Co-dec°*
Sirius	Alpha Canis Majoris	-1.5	-17	-73
Canopus	Alpha Carinae	-0.7	-53	-37
	Alpha Centauri	-0.3	-61	-29
Arcturus	Alpha Boötis	-0.0	+19	+71
Vega	Alpha Lyrae	0.0	+39	+51
Capella	Alpha Aurigae	0.1	+46	+44
Rigel	Beta Orionis	0.1	-08	-82
Procyon	Alpha Canis Minoris	0.4	+02	+88
Achernar	Alpha Eridani	0.5	-57	-33
Betelgeux	Alpha Orionis	var.	+07	+83
Agena	Beta Centauri	0.6	=60	-30
Altair	Alpha Aquilae	0.8	+09	+81
Acrux	Alpha Crucis	0.8	-63	-27
Aldebaran	Alpha Tauri	0.8	+17	+73
Antares	Alpha Scorpii	1.0	-26	-64
Spica	Alpha Virginis	1.0	-11	-79
Pollux	Beta Geminorum	1.1	+28	+62
Fomalhaut	Alpha Pisces Australis	1.2	-30	-60
Deneb	Alpha Cygni	1.2	+45	+45
	Beta Crucis	1.2	-60	-30
Regulus	Alpha Leonis	1.3	+12	+78

Some Other Stars

Shaula	Lambda Scorpii	1.6	-37	-53
	Gamma Crucis	1.6	-57	-33
Alnair	Alpha Gruis	1.7	-47	-43
Kaus Australus	Epsilon Sagittarii	1.8	-34	-56
Regor	Gamma Velorum	1.8	-47	-43
Mirphak	Alpha Persei	1.8	+50	+40
Dubhe	Alpha Ursae Majoris	1.8	+62	+28
Alkaid	Eta Ursae Majoris	1.9	+49	+41
Sargas	Theta Scorpii	1.9	-43	-47
	Alpha Pavonis	1.9	-57	-33
Alphard	Alpha Hydrae	2.0	-09	-81
Polaris	Alpha Ursae Minoris	2.0	+89	+01
Nunki	Sigma Sagittarii	2.0	-26	-64
Mira	Omicron Ceti	var.	-03	-87
Mizar	Zeta Ursae Majoris	2.1	+55	+35
Alpheratz	Alpha Andromedae	2.1	+29	+61
Haratan	Theta Centauri	2.1	-36	-54
Algol	Beta Persei	2.1v	+41	+49
Eltamin	Gamma Draconis	2.2	+51	+39
Suhail Hadar	Zeta Puppis	2.2	-40	-50
Shedir	Alpha Cassiopeiae	2.2v	+57	+33
	Alpha Lupi	2.3	-47	-43
Ankaa	Alpha Phoenicis	2.4	-42	-48
Alderamin	Alpha Cephei	2.4	+62	+28
	Beta Hydri	2.8	-77	-13
Acamar	Theta Eridani	2.9	-40	-50
	Beta Pictoris	3.8	-51	-39
	Sigma Octantis	5.5	-89	-01

Locations

New Zealand	*S.lat.*	*Co-lat.*
Invercargill	46	44
Dunedin	46	44
Christchurch	44	46
Wellington	41	49
Napier	39	51
Auckland	37	53
Australia		
Hobart	43	47
Canberra	35	55
Adelaide	35	55
Sydney	34	56
Perth	32	58
Brisbane	27	63
Darwin	12	78
South Africa		
Cape Town	34	56
Durban	30	60
Johannesburg	26	64
Zimbabwe		
Harare	18	72
Indonesia		
Denpasar	09	81
South America		
Port Stanley	52	38
Buenos Aires	35	55
Montevideo	35	55
Santiago	33	57
Rio de Janeiro	23	67
Iquique	20	70

astronomer named Royer (previously it had been included in Centaurus), and it is not even a true cross; it is shaped more like a kite. It contains four brilliant stars; Alpha Crucis or Acrux (magnitude 0.8), Beta (1.2), Gamma (1.6) and Delta (2.8), while a rather fainter star, Epsilon (3.6) rather spoils the symmetry. I doubt whether anyone will take long to identify it, but in case of any doubt a line is given to it — more or less — by the two Pointers, Alpha and Beta Centauri, which lie near each other and make up a completely unmistakable pair. I have included them in the diagram, even though they belong to the neighbouring constellation.

As we have noted, Alpha, Beta and Delta Crucis are all hot and bluish-white, while Gamma is a red giant of spectral type M. Gamma is only four-tenths of a magnitude fainter than Beta,

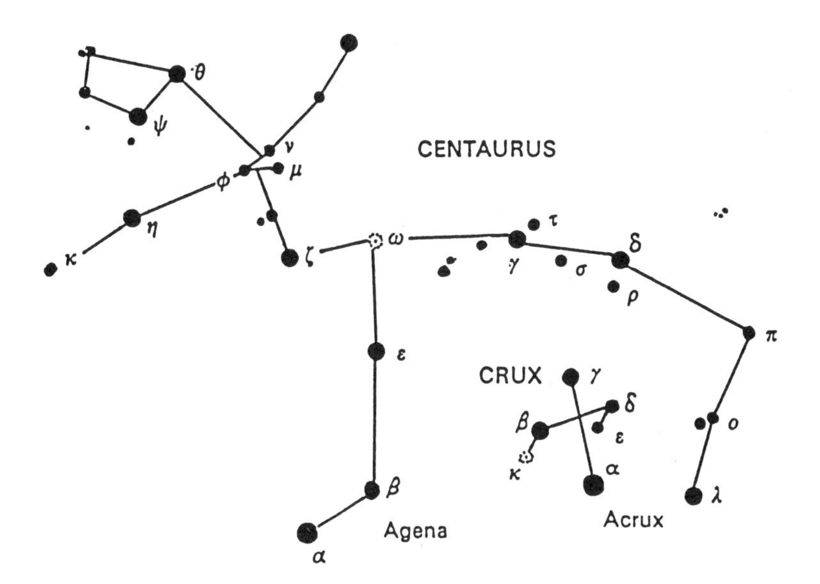

The Southern Cross, with Centaurus.

but for some reason or other Beta is always classed as being of the 'first magnitude' while Gamma is not. A small telescope will show that Alpha Crucis is a wide, easy double; the components are of magnitudes 1.4 and 1.9 respectively, and the separation is 4.4 seconds of arc, while there is a fifth-magnitude star in the same field.

The Milky Way is very rich in Crux, and here too we find the glorious open cluster NGC 4755, round Kappa Crucis, which is nicknamed the Jewel Box. I have already said something about it — the single red supergiant contrasts with the bluish-white of the other main stars in the cluster, three of which form a triangle. The Jewel Box is about 7700 light-years away and is thought to be young, with an age of no more than a few million years. Binoculars give a good view of it, close to Beta Crucis. By sheer coincidence, it lies at the edge of the famous dark nebula of the Coal Sack. There is no real connection, because the Coal Sack is only 500 light-years away and lies in the foreground, so to speak. There are a few stars in front of it, but not many, and at first glance it really does look like 'a hole in the sky'.

Now let us move on to Chart 1 and start using the Cross as a guide to the constellations round the south celestial pole. Our first target is Achernar, in Eridanus (the River). It and the Cross lie on opposite sides of the pole, and are at about the same distance from it, so that when the Cross is high up Achernar is low down, and vice versa. Achernar is of magnitude 0.5, and

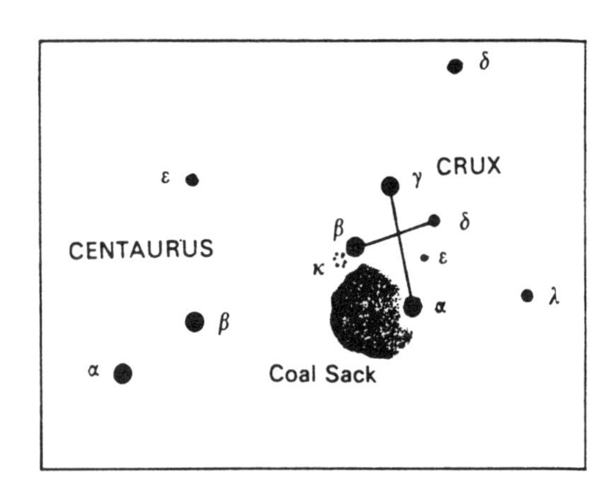

Position of the Coal Sack, the most prominent of the dark nebulae.

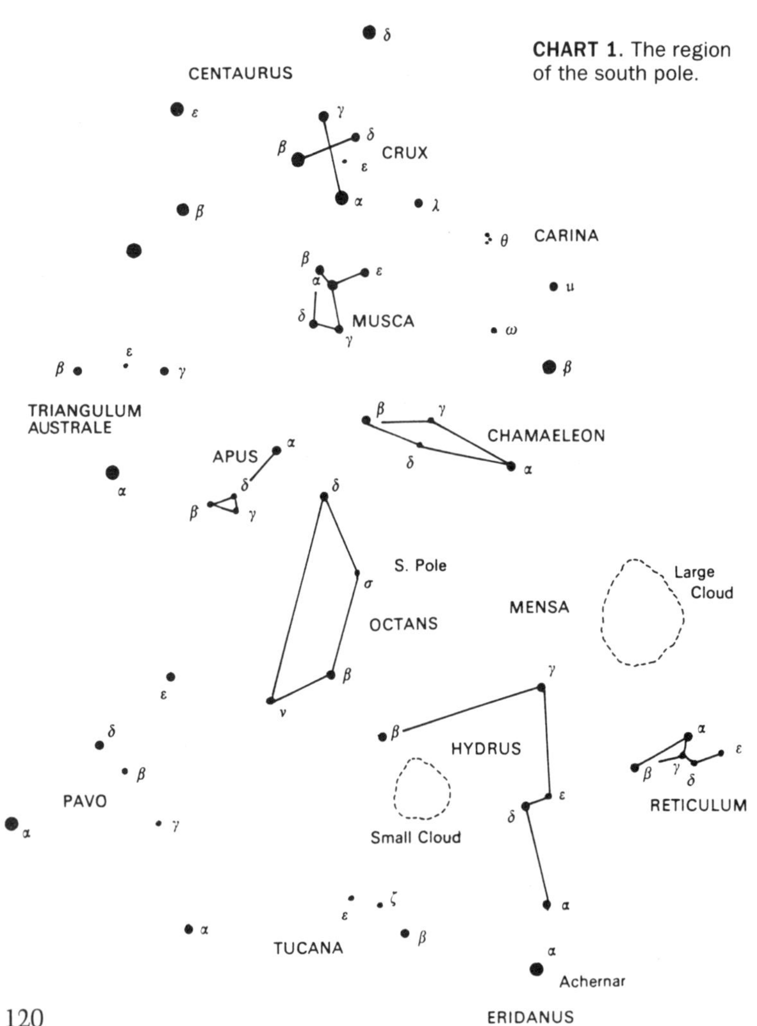

CHART 1. The region of the south pole.

is actually brighter than any of the stars in the Cross; it is hot and white, and 780 times as luminous as the Sun.

It cannot honestly be said that the immediate region of the pole contains much of interest, and there are various faint, undistinguished constellations: Octans (the Octant), Apus (the Bird of Paradise — originally Avis Indica), Chamaeleon (the Chameleon), Mensa (the Table) and Dorado (the Swordfish). Slightly more prominent are Reticulum (the Net), which has at least a distinctive shape, and Musca (the Fly), a condensed little group which is easy to find because it lies next to the Cross. Hydrus (the Little Snake) has three stars of the third magnitude, and one of these, Beta Hydri, is the nearest reasonably bright star to the pole itself. During periods of moonlight, the whole area will appear more or less blank.

However, we do at least have the unique Clouds of Magellan, which are superb objects. They contain objects of all kinds. There is, for instance, the Tarantula Nebula in the Large Cloud, which is far larger than the Orion Nebula, and would cast shadows if it lay within a few hundred light-years of us. Also in the Large Cloud is a particularly luminous star, S Doradûs, which outshines our Sun by at least a million times, and yet is too far away to be seen with the naked eye!

Many people will be interested in locating the south pole star, Sigma Octantis, which is only of magnitude 5.5, and is none too easy to see with the naked eye unless the sky is really dark and clear. Octans itself contains only one star above the fourth magnitude (Nu Octantis, 3.8) and has no obvious shape, which makes matters even more troublesome. My method involves using wide-field binoculars — say 7 x 50 — and I have found that it works. First identify Alpha Apodis, which lies in the direction of a line through Alpha Centauri and Alpha Circini. In the same field as Alpha Apodis are two faint stars, Epsilon

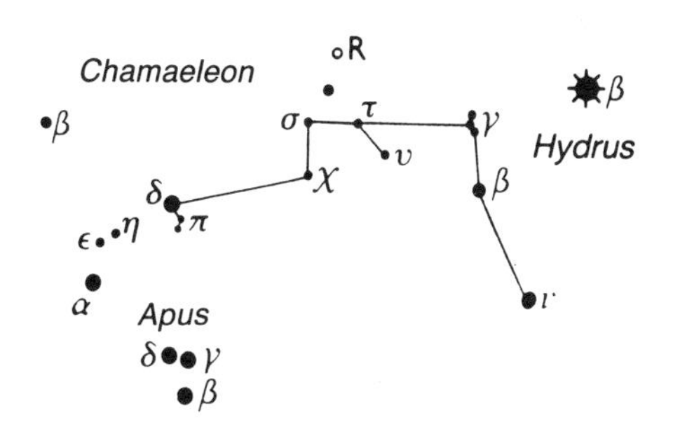

My method of identifying Sigma Octantis.

Apodis (5.2) and Eta Apodis (5.0). These point straight to the orange Delta Octantis (4.3) which has two dim stars, Pi^1 and Pi^2 Octantis, close beside it. Now put Delta in the edge of the binocular field, and continue the line from Apus. Chi Octantis (5.2) will be on the far side of the field. Centre it, and then you will see two more stars of about the same brightness, Sigma and Tau. These three are in the same field, and make up a triangle; the south polar star, Sigma, is the second in order from Delta. This may sound complicated, but I hope that the diagram will help.

Sigma Octantis is 120 light-years away, and is only six times as powerful as our Sun, so that it is by no means the equal of the northern hemisphere's Polaris. Yet things will not always be the same. The polar points of the sky shift very slowly, because of an effect known as precession, and in around 12,000 years from now the south pole star will be the brilliant Canopus. Meantime, we must make do with the puny Sigma Octantis!

Chapter 18

Stars Of Summer Evenings

Now that we have broken the celestial ice, and have dealt with a few of the groups which are nearly always on view, it is time to begin our seasonal review of the year. The system I have adopted is to give two basic charts for each season, one showing the aspect of the sky when looking south and the other the aspect when looking north. If you are in any doubt about the points of the compass, locate the south celestial pole by using the longer axis of the Cross, and stand with your back to it. You will then be looking north, with east to your right and west to your left.

The hemispherical charts on page 124 do not pretend to be anything but basic. They are not of high accuracy, and on this projection cannot be so, because there is bound to be a certain amount of distortion. But they will, I hope, meet our immediate requirements, and the supplementary charts go into more detail.

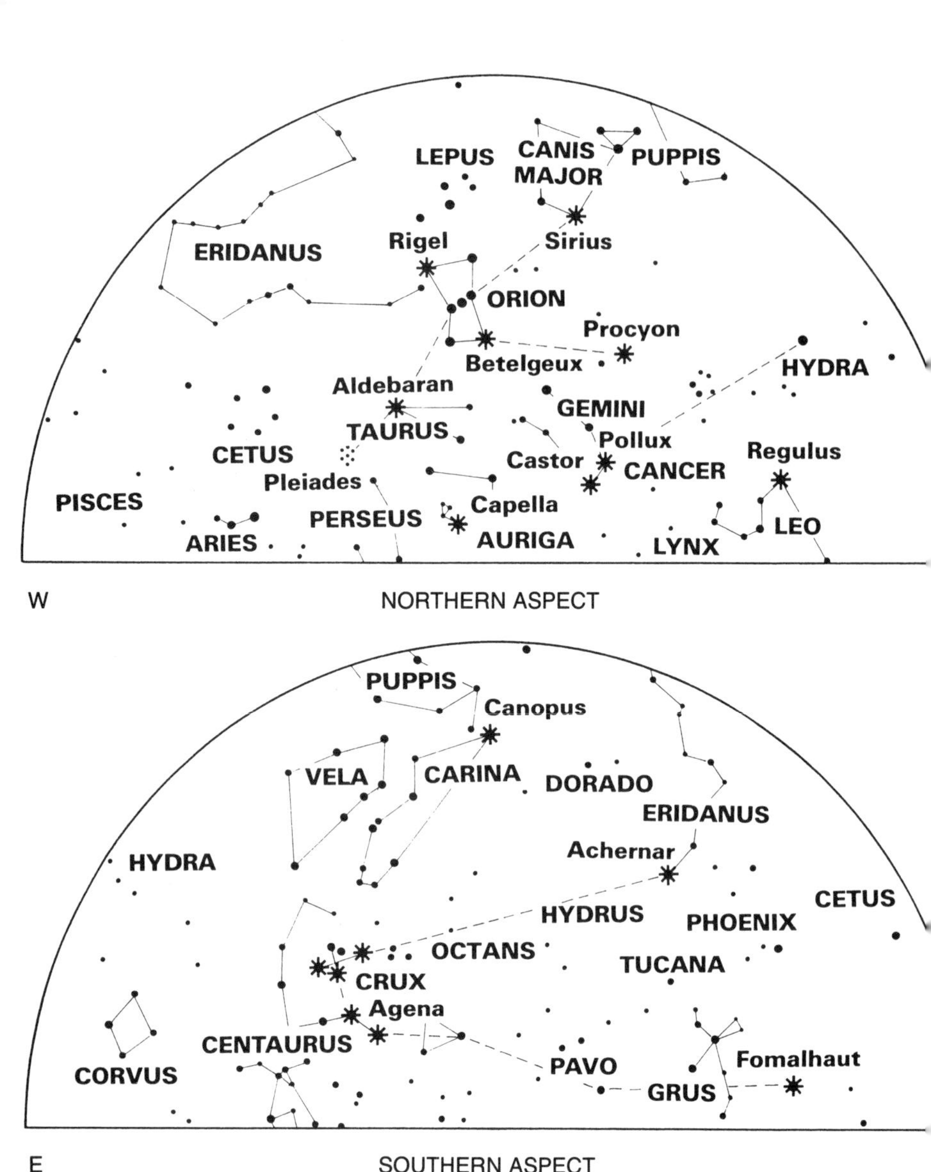

The evening sky in summer.

They have been drawn for around 10 p.m. in mid-January, but they can be used for other times of night at other seasons. The difference is about two hours per month, so that these maps also apply to 8 p.m. in mid-February, 6 p.m. in mid-March and so on; or, for that matter, to midnight in mid-December, 2 a.m. in mid-November and 4 a.m. in mid-October.

I have selected January partly because it is the beginning of the year, and partly because both our main guides, Orion and the Cross, are on view. Orion, indeed, is not far from the zenith or overhead point, so let us deal with it first. It is shown to a larger scale in the separate diagram, and is very easy to recognise, because its seven main stars are so brilliant and make up so characteristic a pattern. Beta (Rigel) and Alpha (Betelgeux) are among the brightest of the first-magnitude stars; then follow Gamma (Bellatrix), Epsilon (Alnilam), Zeta (Alnitak), Kappa (Saiph) and Delta (Mintaka), all of which are between magnitude 1.6 and 2.2.

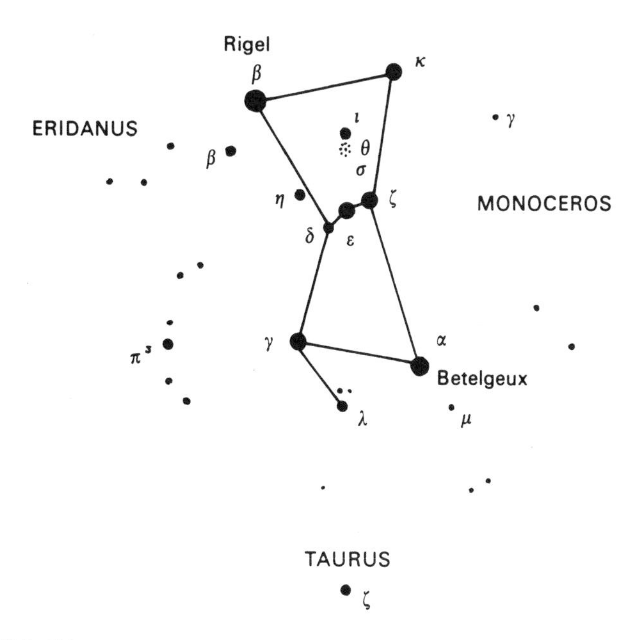

CHART 2. Orion.

Of the two leaders, Rigel is the brighter, and should logically have been lettered Alpha rather than Beta. It is a true celestial searchlight, at least 60,000 times more powerful than the Sun. Its distance is about 900 light-years, so that we are now seeing it as it used to be when William the Conqueror ruled England. When seen through binoculars it is magnificent, and it has a seventh-magnitude companion at a separation of between 9 and 10 seconds of arc.

Betelgeux is as different as it could possibly be. Here we have a red supergiant, well past its prime of life and experiencing its last period of glory before disaster overtakes it. In sheer power

Orion and his retinue.

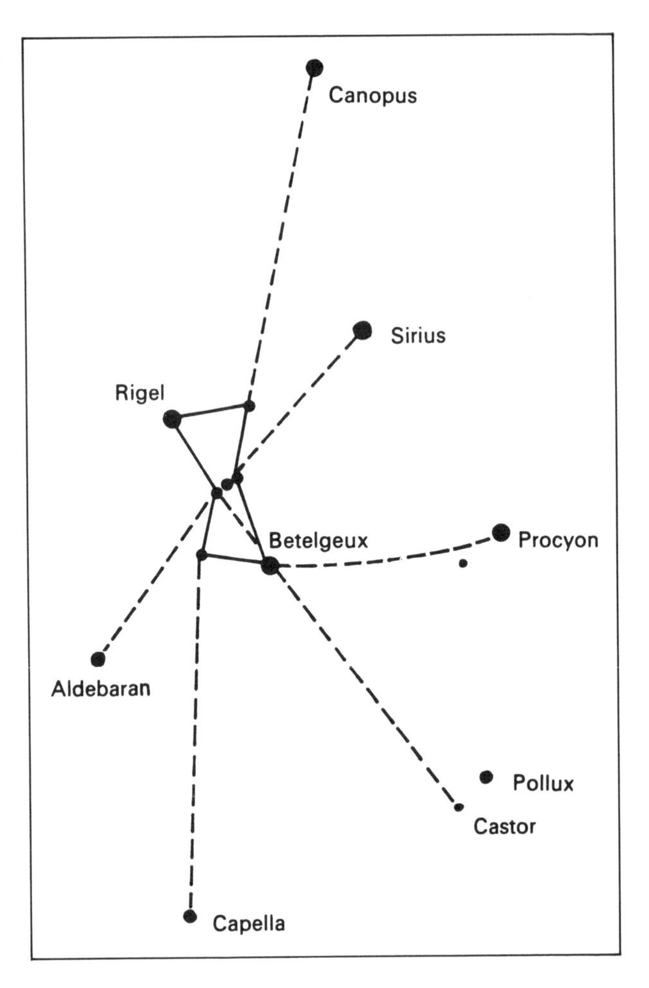

it cannot compare with Rigel, even though it could match 15,000 Suns, and its surface is relatively cool — hence its orange-red colour — but what it lacks in temperature it makes up for in sheer size. Its diameter is 400 million km, so that it could engulf the whole orbit of the Earth round the Sun.

Betelgeux is decidedly variable. It is an unstable star; it swells and shrinks, changing its output as it does so. The magnitude is generally about 0.5, but there are times when it rises almost to equality with Rigel, while at minimum it is not much brighter than Aldebaran in Taurus. It is classed as a semi-regular variable, with a period of about 5.7 years, but this period is very rough indeed. Its fluctuations are slow, but can be followed with the naked eye. The method is to compare Betelgeux with other bright stars, mainly Procyon in Canis

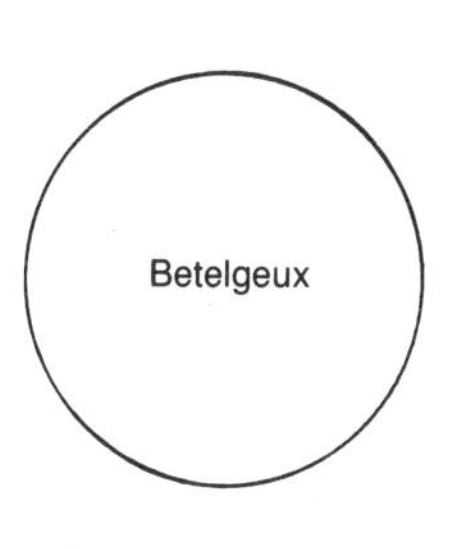

Betelgeux and the Sun compared.

Minor (0.4), Aldebaran (0.8) and, of course, Rigel (0.1). Some allowance must be made for what is termed 'extinction'. A star which is low on the horizon will be dimmed because more of the incoming light is absorbed by our atmosphere. In making comparisons, always try to select a comparison star which is at about the same altitude as the variable.

According to the Cambridge catalogue, Betelgeux is 310 light-years away. This means that it is only one-third as far from us as Rigel, and there is absolutely no real connection between the two. Modern techniques have made it possible to identify a certain amount of surface detail on it, though ordinary telescopes show it only as a speck of light.

The other main stars in Orion are very hot and bluish-white; all are very luminous — Saiph seems to be about 50,000 times as powerful as the Sun. Look next at the Sword, which extends upward from the three stars marking the Hunter's Belt (Alnitak, Alnilam and Mintaka). Here we find the Great Nebula, M.42, which I have already described; a small telescope will show the four main components of the Trapezium, Theta Orionis, which make the nebula shine. Closer to Alnitak is a small dark gas-patch, known as the Horse's Head from its outward resemblance to the head of a knight in chess. It is not particularly difficult to photograph, but is remarkably elusive telescopically. The Milky Way flows close to Orion, and rich star-fields abound.

The Hunter's retinue is shown in the next diagram. Pride of place must go to Sirius in Canis Major (the Great Dog), which shines as the most brilliant star in the whole of the sky. To locate it, start at the Belt and follow the line upward; but no guide should be needed, because Sirius stands out at once.

It is one of our closest stellar neighbours, at a distance of only 8.6 light-years. It is 26 times as luminous as the Sun. This is puny in comparison with Rigel, and Sirius is not so imposing as it looks. In colour it is pure white, but when low over the horizon it seems to flash all colours of the rainbow; even when high up, as on summer evenings, it still twinkles noticeably. In fact, star-twinkling is due entirely to the Earth's dirty, unsteady atmosphere, and has nothing directly to do with the stars themselves. Go beyond the top of the atmosphere (to the Moon, for example!) and the stars will appear as hard, steely points, without the slightest tendency to shimmer and flicker. From Earth, Sirius is the supreme 'twinkler', merely because it is so brilliant. (*En passant*, it is often said that planets do not twinkle, because they show up as tiny disks instead of point sources. This is not entirely true, but certainly a planet twinkles much less than a star at the same altitude.)

Sirius is not a lone traveller in space. As we have noted, it has a faint white dwarf companion, as massive as the Sun but no larger than a planet such as Uranus; the orbital period is 50 years. Because Sirius is often called the Dog-Star, the Companion is nicknamed the Pup, but it is a very substantial pup indeed.

There is a minor mystery associated with Sirius. Some of the old astronomers of Classical times described it as being red.

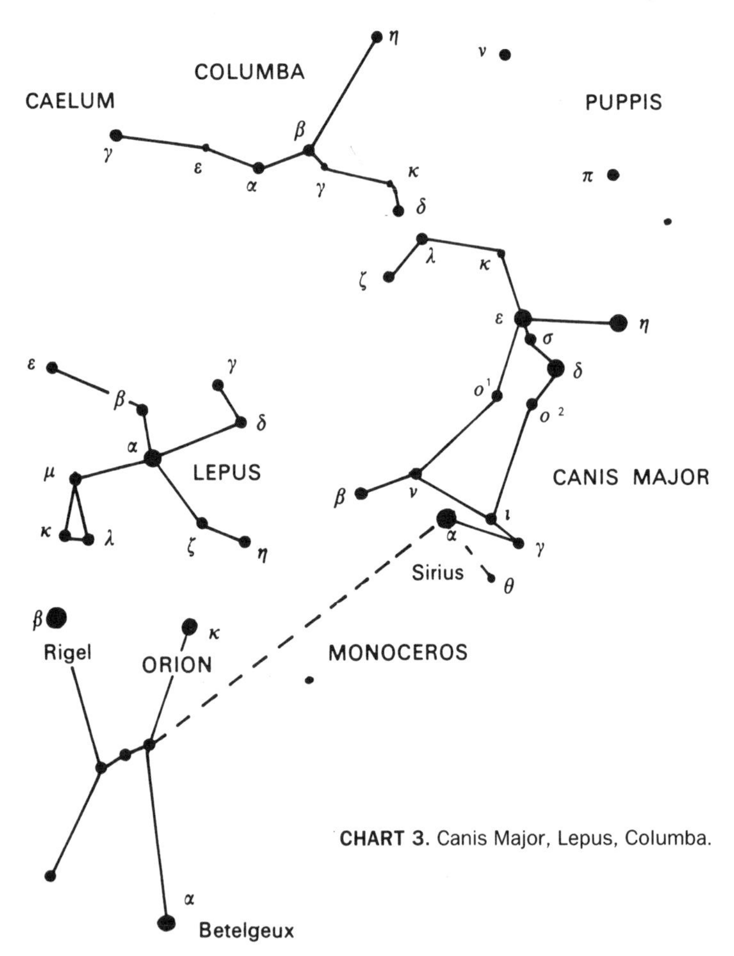

CHART 3. Canis Major, Lepus, Columba.

This is certainly not true today, despite the twinkling, and it is not likely that there has been any real alteration, because Sirius is a perfectly stable Main Sequence star. Suggestions that the Pup may have been a red giant not so long ago can be ruled out, mainly because the time-scale is all wrong. We have to conclude that there has been some error in observation or interpretation; all the same, it is decidedly curious.

Canis Major contains several bright stars as well as Sirius; Epsilon (Adhara) magnitude 1.5, Delta (Wezea) 1.9, Beta (Mirzam) 2.0 and Eta (Aludra) 2.4. All are very remote and luminous, so that they far outshine Sirius. Wezea may be 130,000 times as powerful as the Sun, but is over 300 light-years away. There is also a fine open cluster, M.41, which lies close to Sirius. It is visible with the naked eye as a hazy patch, and binoculars show it well. Canis Major is shown in Chart 3, together with two of its smaller neighbours, Columba (the Dove) and Lepus (the Hare), neither of which is of particular note.

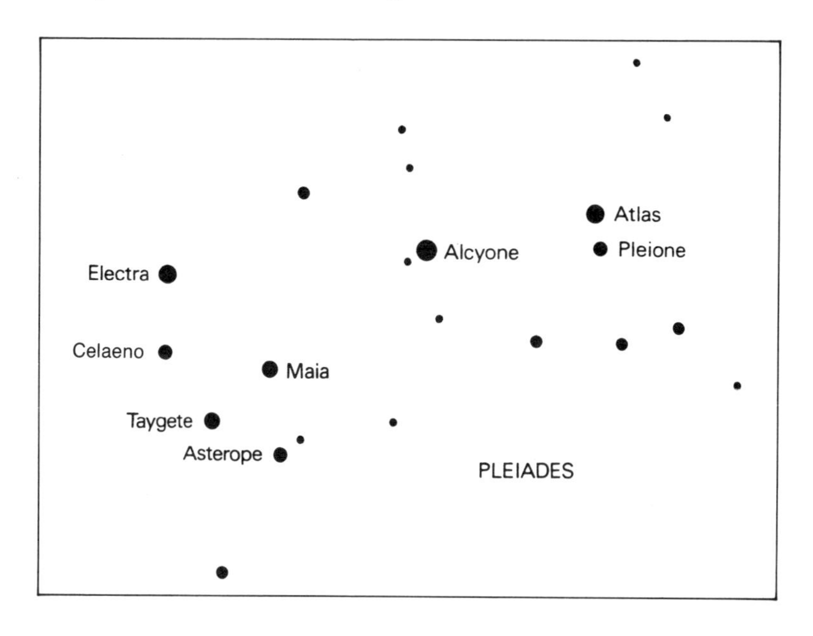

THE PLEIADES (M.45), in Taurus (see Chart 4). The brightest stars, with their magnitudes, are:

Alcyone	2.9
Atlas	3.6
Electra	3.7
Maia	3.9
Merope	4.2
Taygete	4.3
Pleione	5.1 (variable)
Celaeno	5.5
Asterope	5.8

Now let us return to Orion's Belt, and follow the line downward. This time we will reach an orange-red first-magnitude star, Aldebaran, the 'Eye' of Taurus (the Bull). Superficially it looks very much like Betelgeux, though it is neither so large, so remote or so powerful. Extending away from it is the V-shaped cluster of the Hyades, and on the far side of Orion (Chart 4) we come to the Pleiades or Seven Sisters, which were of such significance to the Maori. On the next clear night, count the number of Pleiads you can see with the naked eye. The record, held by a last-century German astronomer named Eduard Heis, is said to be 19.

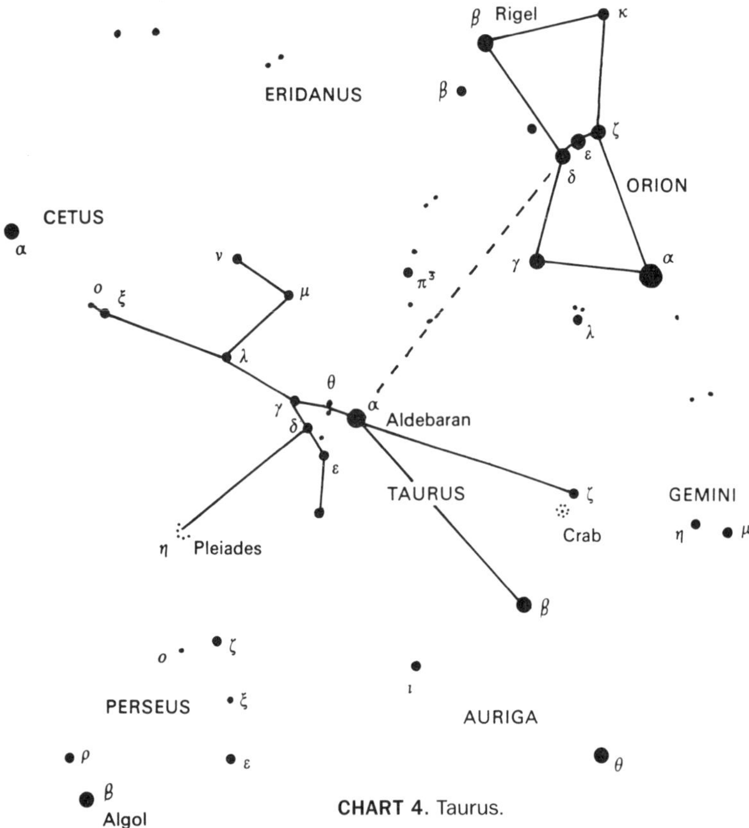

CHART 4. Taurus.

Taurus is not a constellation with a distinctive shape, but it does contain one other bright star, Beta Tauri or Alnath (1.6). Alnath was formerly included with Auriga, as Gamma Aurigae, and there seems to be obvious reason for its free transfer to the Bull. Lambda Tauri, also shown on Chart 4, is an Algol-type eclipsing binary with a magnitude range of 3.3 to 4.2 and a period of 3.9 days. It can be found by using the V-formation of the

Hyades as a sort of 'arrow-head'; useful comparison stars are Gamma (3.6), Xi (3.7) and Mu (4.3).

The Crab Nebula, known to be the remnant of the supernova of 1054, lies close to the third-magnitude Zeta Tauri or Alheka. It can just be seen with powerful binoculars. Outwardly it may look dim and uninteresting, but to astronomers it is one of the most important objects in the sky, because it radiates over almost the whole range of wavelengths. One comment which has been widely repeated is that 'there are two kinds of astronomy; the astronomy of the Crab Nebula, and the astronomy of everything else'! It was No.1 in Messier's catalogue; its nickname of the Crab was bestowed on it in 1845 by the Earl of Rosse, who examined it with his great home-made 183-cm reflector at Birr Castle in Ireland.

Next in our catalogue of Orion's retinue comes Auriga, the Charioteer (Chart 5). It is easy to locate, because it contains one exceptionally brilliant star, Capella. In case of difficulty, take a guide-line from Mintaka, in the Hunter's Belt, through Bellatrix. Auriga can attain a reasonable height during summer evenings from most of Australia and South Africa — look for it in the north — but from New Zealand it is always inconveniently low, and from the southernmost part of the South Island Capella never rises at all.

Capella is a yellow star. Its surface temperature is much the same as that of the Sun, but whereas the Sun is officially ranked as a dwarf, Capella is a giant, with a luminosity around 150 times that of the Sun. It is actually a very close binary, but the components are only about 110 million km apart, and in an ordinary telescope Capella looks single. Its distance from us is 42 light-years.

Adjoining Capella is a triangle of three fainter stars, Epsilon, Eta and Zeta Aurigae, known collectively as the Haedi, or Kids (Capella has been known as the 'She-Goat'). One member of the trio, Eta, is an ordinary bluish-white star, 400 times as luminous as the Sun, but the other two are remarkable objects. Both are eclipsing binaries, though they are not in the least alike. Zeta (Sadatoni) has a range of from magnitude 3.7 to 4.2; eclipses occur every 972 days. The main star is an orange K-type supergiant, whose colour is obvious in binoculars; the companion is a hot B-type star. The distance from us is nearly 700 light-years.

Epsilon Aurigae (seldom referred to now by its old proper name of Almaaz) is even more unusual. Here we have a particularly luminous supergiant, perhaps well over 200,000 times as powerful as the Sun, lying at a distance of 4600 light-years. The eclipsing companion has never been seen, and if it did not periodically pass in front of the supergiant we would

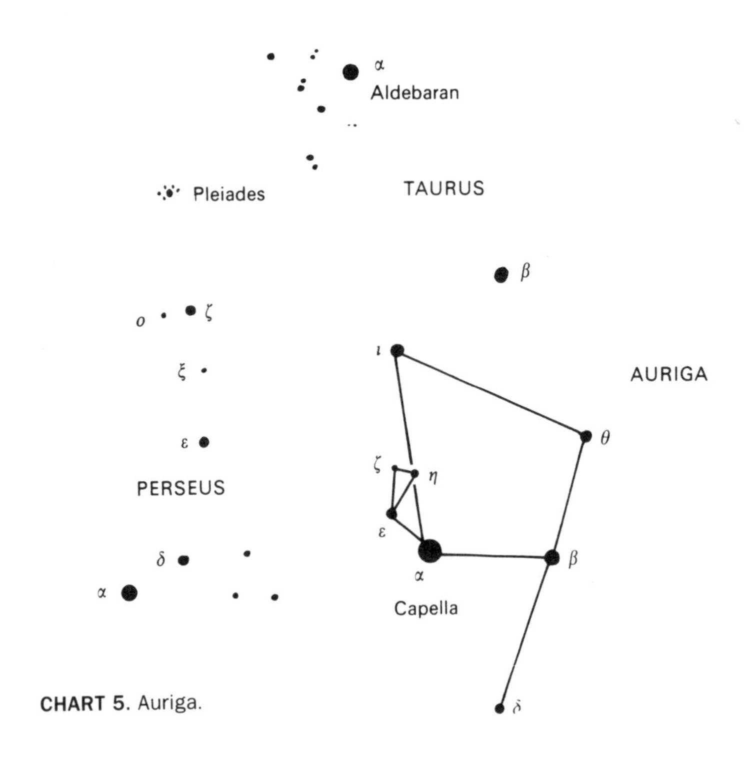

CHART 5. Auriga.

know nothing about it. We are not even sure of its nature, though most astronomers think that it must be a small, hot star, surrounded by a large opaque 'cloud'. The period is 27 years, and eclipses last for a very long time. The extreme range is from magnitude 2.9 to 3.8, and Eta makes a convenient comparison star. The last eclipse began on 22 July 1982 and did not end until 25 June 1984, though it was total only for a year (January 1983 to January 1984). Astronomers will be eagerly waiting for the next eclipse, due to start in the year 2009.

The rest of Auriga is fairly prominent; Beta (Menkarlina) is of the second magnitude, Theta and Iota between 2 and 3. There are also three fine open clusters, Messier 36, 37 and 38, all of which are binocular objects — and in Auriga, the Milky Way is very rich.

The last two members of Orion's retinue are Gemini (the Twins) and Canis Minor (the Little Dog). Gemini is marked by two bright stars, Pollux (1.1) and Castor (1.6), which lie side by side; you can find them by using Rigel and Betelgeux as pointers. In mythology they were twin boys; Pollux was immortal, while Castor was not. When Castor was killed in battle, Pollux was pleased to be allowed to share his immortality with his brother,

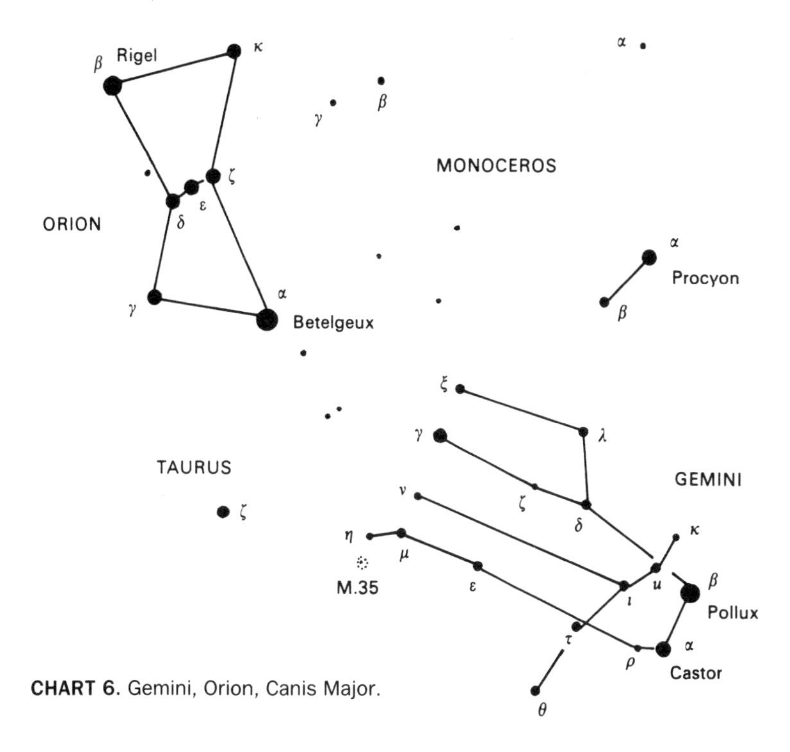

CHART 6. Gemini, Orion, Canis Major.

and both were transferred to the heavens. Though lettered Beta, Pollux is the brighter of the two by half a magnitude.

Castor and Pollux are not genuinely associated; Castor is 46 light-years away from us, Pollux only 36 light-years. Pollux is a single star of type K, clearly orange in colour and 60 times as luminous as the Sun; Castor is white, and is a multiple system. It is a fine double; the components are of magnitudes 1.9 and 2.9 respectively, and the separation is between 2 and 3 seconds of arc. The orbital period is 420 years. Each component is again double, though too close to be split with an ordinary telescope, and there is a third, faint companion which also is binary, so that altogether Castor consists of six suns, four bright and two dim.

The form of Gemini is shown in Chart 6. The Milky Way flows through it, so that there are many rich star-fields, and there are two interesting variable stars, the semi-regular Eta or Propus (range 3.2 to 3.9, rough period 233 days) and Zeta or Mekbuda, which is a typical Cepheid with a range of from 3.7 to 4.1 and a period of 10.15 days. For these variables you may like to have the magnitudes of some comparison stars: Mu or Tejat (2.9, reddish); Nu (4.1); Xi (3.5); Lambda (3.6) and Delta (3.5). There is also a splendid open cluster, Messier 35, close to Mu and the variable Eta. It is on the fringe of naked-eye visibility, and binoculars show dozens of stars in it, most of which are hot and white.

Canis Minor contains one brilliant star, Procyon (0.4). Like Sirius, it is one of our nearer neighbours, at 11.4 light-years, and it too has a dim white dwarf companion. Canis Minor contains nothing else of note, so I have included it in the map with Gemini (p. 133). The area between Procyon, Sirius and Betelgeux is filled by a large constellation, Monoceros (the Unicorn), which is rich in faint stars and lovely fields, but contains no bright star.

If you extend a line from Rigel through Aldebaran, and continue it as far north as you can, you will come to Perseus. It contains one second-magnitude star, Mirphak, but is always very low because of its high northern declination, and from Invercargill it remains below the horizon. The most interesting object in Perseus is the prototype eclipsing binary Algol; the period is 2.87 days, and the range from magnitude 2.3 to 3.5. It remains at maximum for most of the time. When it fades, it takes several hours to fall to minimum, and after remaining dim for 20 minutes it brightens slowly up again. Many other eclipsing

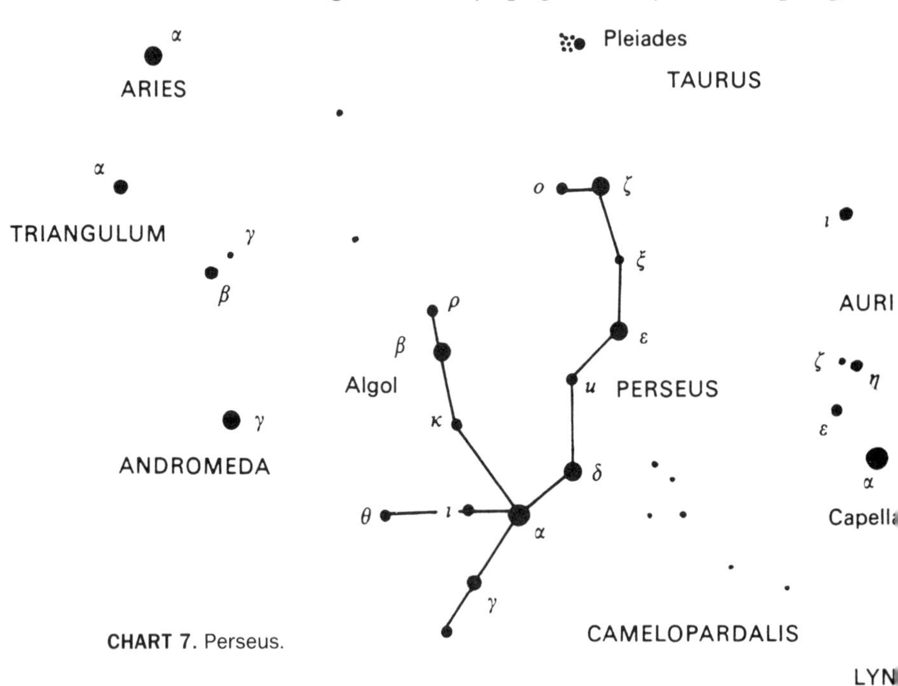

CHART 7. Perseus.

binaries of the same type are known but only Lambda Tauri and Zeta Phoenicis are easily visible with the naked eye apart from Algol itself.

Of Algol's two components, the brighter is a B-type star about 100 times as luminous as the Sun; the eclipsing companion is larger but less powerful, and the eclipse is not quite total. When the fainter star passes in front of the brighter there is a very small dip in magnitude, but you will not notice this with the naked eye. Incidentally, the explanation of Algol's behaviour was discovered in 1783 by a most unusual astronomer, John Goodricke, who was deaf and dumb, and who died at the early age of 21.

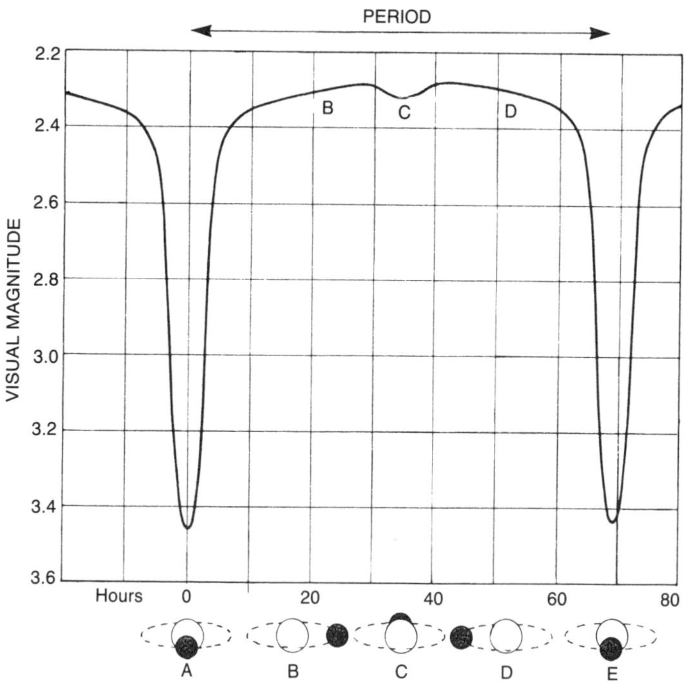

THE VARIATIONS OF ALGOL. Principal minimum occurs when the brighter component is partially hidden by the fainter member of the pair. The secondary minimum, when the fainter component is partly hidden, is too slight to be noticed with the naked eye. The southern Zeta Phoenicis is a star of the same type of Algol.

We need waste no time on the extremely dull northern constellations of Lynx (the Lynx) and Camelopardalis (the Giraffe). In the north-east Leo, the Lion, is rising, but will be better placed in the autumn; mention must be made of the second-magnitude Alphard or Alpha Hydrae, which is yet another red giant, and may be found by using Castor and Pollux as pointers, and we can postpone, too, any discussion of Cetus,

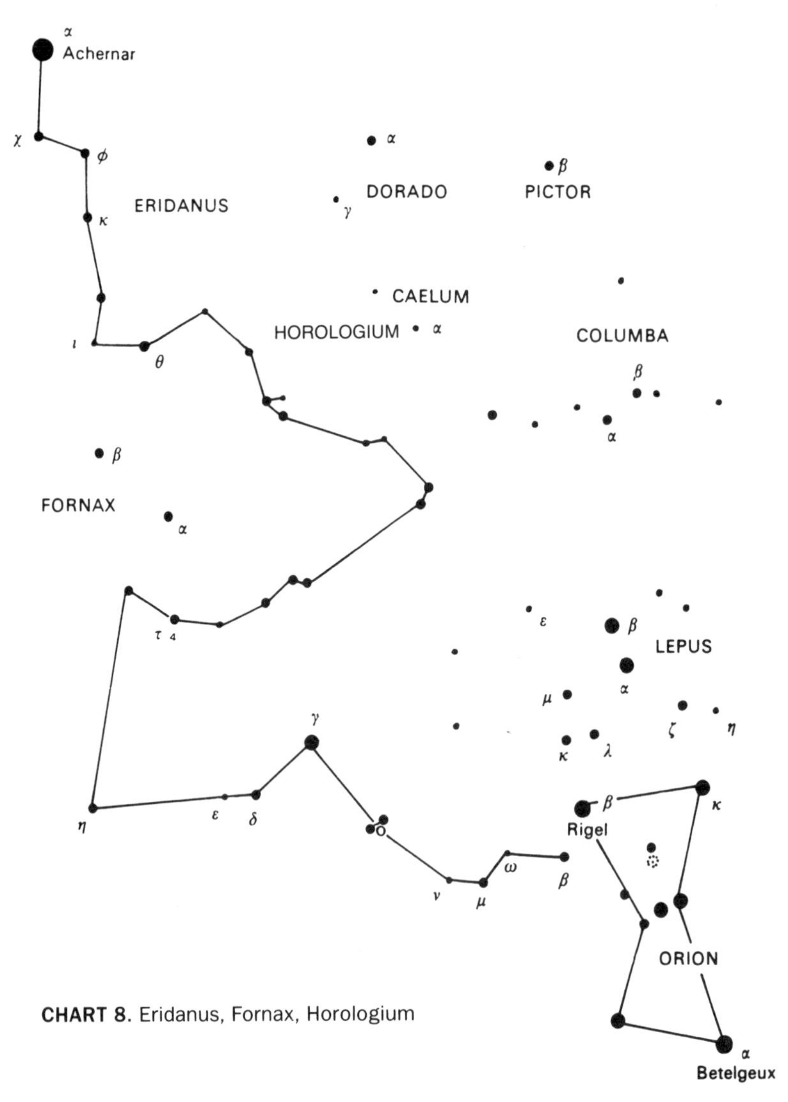

CHART 8. Eridanus, Fornax, Horologium

Pisces and the other constellations sinking in the north-west. Meanwhile, let us turn back to the southern aspect of the sky. The Cross is rather low, and the Pointers, Alpha and Beta Centauri, are not far above the horizon. (They actually set from Brisbane and Johannesburg, but are circumpolar from Perth, Durban or anywhere further south, and of course from the whole of New Zealand.) Achernar can be located by using the longer axis of the Cross, as shown earlier. The long constellation of Eridanus extends from the far south right up to Orion, and seems to merit a separate map, which I have given in Chart 8. Theta Eridani, or Acamar, is a fine, easy double; the components are of magnitudes 3.4 and 4.5, and the separation is over 8 seconds of arc. Both components are hot and white, with A-type spectra. Some early catalogues ranked Acamar as being of the first magnitude, but the combined magnitude is now only 2.9, and there may have been some confusion with Achernar, particularly as Acamar is known as 'the Last in the River' even though it is not at the end of the line. Epsilon Eridani is much further north, and only of magnitude 3.7, but it is less than 11 light-years away, and although it is smaller and less powerful than the Sun it is sufficiently 'solar' to be regarded as a possible planetary centre. It was one of two stars to be studied by Drake and his team in their original hunt for extra-terrestrial intelligence, though the results were, alas, negative.

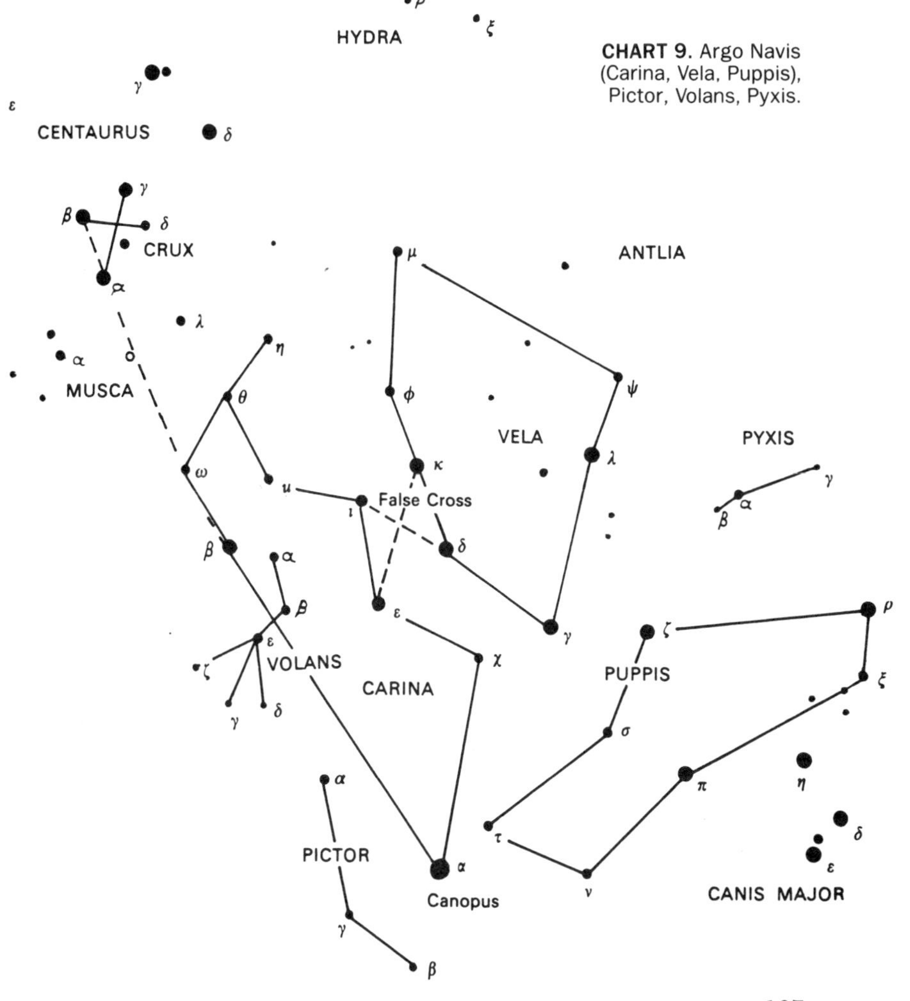

CHART 9. Argo Navis (Carina, Vela, Puppis), Pictor, Volans, Pyxis.

The high south is dominated by Argo Navis, the Ship Argo, which contains the brilliant Canopus. Argo has the dubious distinction of being the only constellation which has been officially chopped up because it was too large and unwieldy. In mythology it was the ship which carried Jason and his companions in their rather unprincipled quest for the Golden Fleece. Modern astronomers have divided it into Carina (the Keel), Vela (the Sails) and Puppis (the Poop), so that Canopus, which used to be known as Alpha Argûs, has become Alpha Carinae. In Chart 9 (p. 137), I have shown the whole of the old Argo, together with the adjacent groups of Pyxis (the Mariner's Compass), Volans (the Flying Fish) and Pictor (the Painter).

How to find Canopus, using Orion. (Not that any guide is really necessary!)

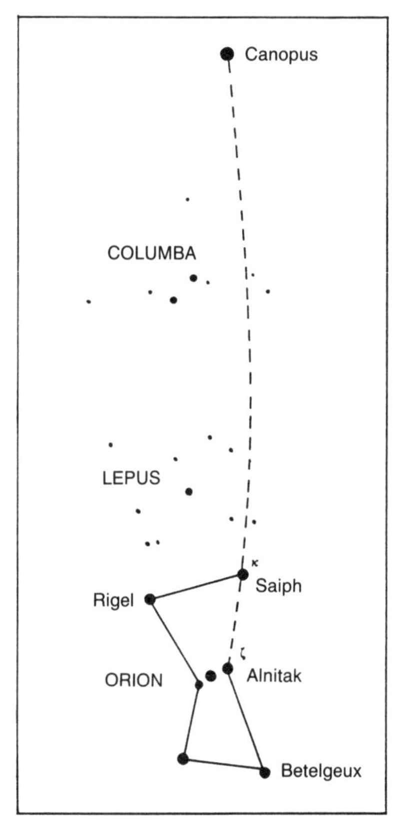

For sheer brilliance, Canopus is inferior only to Sirius; its magnitude is -0.7. To identify it, extend a line from Alnitak in Orion's Belt through Saiph, passing through Lepus and Columba and finally arriving at Canopus. The method is shown in the diagram. Unfortunately the hemispherical charts do not make the relative positions of Canopus and Orion clear (this is one of their weaknesses), but the extra map on this page should prevent any confusion. And in any case, Canopus is much too bright to be overlooked. It has an F-type spectrum, and in theory ought to be slightly yellowish, but to me it appears pure white, and I have never yet met anybody who has claimed to see a yellowish tinge to it. It is a cosmic searchlight, perhaps 200,000 times as luminous as the Sun, and squandering its 'fuel' at a prodigious rate. It is circumpolar from New Zealand, though from parts of Australia and South Africa it sinks briefly below the horizon during its 24-hour circuit of the pole.

The richest areas of Argo are in Carina and Vela. Beta Carinae or Miaplacidus (magnitude 1.7) can be located by a slightly curved line from Beta and Alpha Crucis. If sufficiently prolonged, the same line will eventually lead you to Canopus.

Note also the False Cross, made up of four brightish stars; Kappa and Delta Velorum (magnitudes 2.5 and 2.0 respectively) and Iota and Epsilon Carinae (2.2 and 1.9). Unwary viewers have often mistaken it for the real Cross, and this is understandable, because the shapes are the same; but the False Cross is larger, less brilliant and more symmetrical. Here too there is an interesting colour contrast; three members of the False Cross are bluish-white, while the fourth and brightest, Epsilon Carinae, is orange.

The False Cross can be used to find a fine open cluster, IC 2602, round Theta Carinae. Use Delta Velorum and Iota Carinae as pointers; Chart 9 (p. 137) shows the relative positions. And not far from Theta is the strange, erratic variable Eta Carinae, which, as we have noted, was once even more brilliant than Canopus, but is now just below naked-eye visibility. Binoculars are powerful enough to show its colour, and telescopically I always describe it as an orange 'blob', quite unlike a normal star. At its peak it was perhaps 6 million times more powerful than the Sun, and this is still true today, though its brilliance is dimmed by intervening nebulosity; in infra-red it is still one of the strongest sources in the sky. The Eta Carinae nebula is visible with the naked eye, and is a magnificent sight in a telescope.

It is always worth keeping a watch on Eta Carinae, because it may flare up again at any time. It is highly evolved and unstable, so that eventually it is almost certain to 'go supernova'. We know of nothing else in the sky quite like it.

Adjoining Carina is Pictor, the Painter (originally Equuleus Pictoris, the Painter's Easel — many of these older, cumbersome names have been shortened). Beta Pictoris, of magnitude 3.8, is of special interest because of its possible planetary system. (I have described this earlier.) However, to all appearances Beta Pictoris looks like an ordinary white star.

Have you followed me thus far? I hope so; and of course you can always select your own 'pointers'. There are many alignments. For instance, an alternative way of finding the Twins, Castor and Pollux, is to make use of Canopus and Sirius, as shown here. Remember, too, that the stars I have described in this chapter are not only visible in the summer. Many of them will also be found on the charts for other seasons, and New Zealanders, at least, never lose sight of the best-known of all constellations, the beloved Southern Cross.

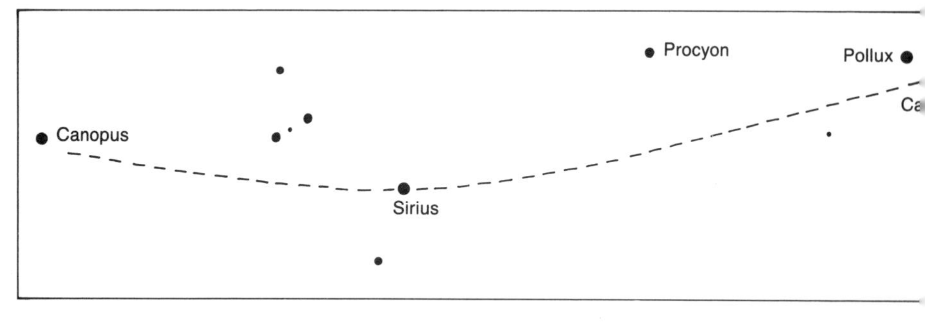

How to find the Twins, using Sirius and Canopus.

Chapter 19

Stars Of Autumn Evenings

Charts 10 and 11 (p. 143) show the aspect of the sky at 10 p.m. in mid-April; the same view applies to 8 p.m. in mid-May or midnight in mid-March. However, we do have to allow for latitude differences. In particular I have not shown Ursa Major, the Great Bear, as from New Zealand it can never be seen at all. From Cape Town the only one of the seven stars which ever rises is the southernmost, Eta Ursae Majoris (Alkaid). From Perth only part of the group can be seen, though all seven main stars do rise briefly from Johannesburg. Nobody in Australia or South Africa is going to appreciate the eminence of the Bear, and for this reason I do not propose to say more about it here, but if you happen to go to Europe, do not forget to look upward and admire it. It may be less brilliant than the Southern Cross, but at least it is a great deal larger!

On our seasonal charts, Orion has almost disappeared, though Sirius is still high up and is dazzlingly conspicuous.

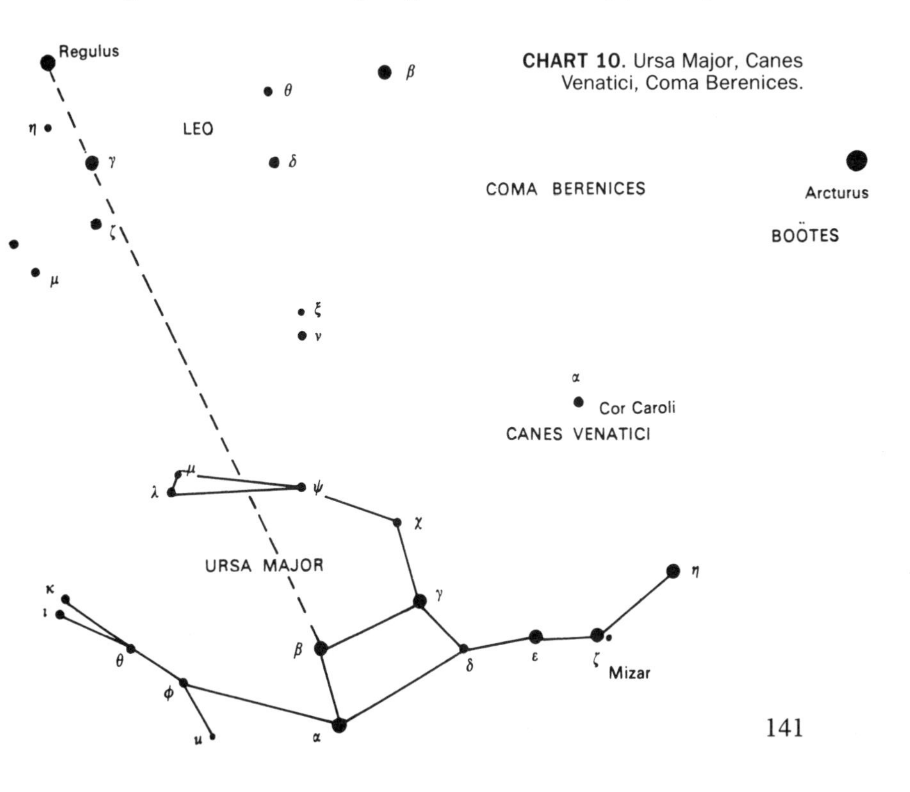

CHART 10. Ursa Major, Canes Venatici, Coma Berenices.

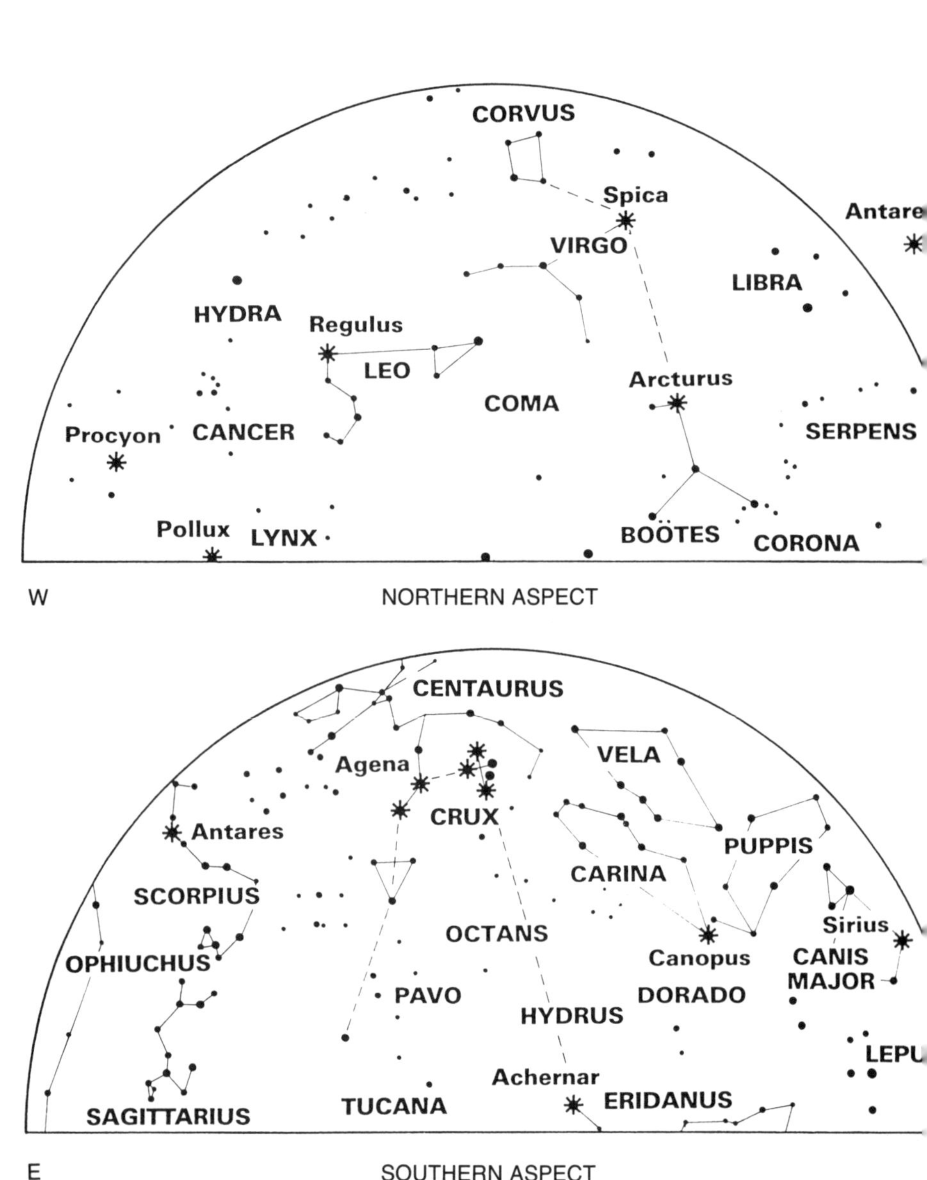

The evening sky in autumn.

Capella has vanished, and so has most of Taurus. However, another of the main northern-hemisphere groups has attained its peak altitude and is easy to locate. This is Leo, the Lion — in mythology, the fearsome Nemaean Lion which met an untimely fate at the hands of the hero Hercules.

When Orion is available, a suitable way to identify Leo is to use Bellatrix (Gamma Orionis) and Procyon as pointers (Chart 11). During evenings in autumn Leo is quite high in the north, and is recognizable both because of its distinctive shape and because of its one first-magnitude star, Regulus. To be candid, Regulus is only of magnitude 1.3, and is not nearly so striking as Procyon, but it is bright enough to stand out. It is white, and is 130 times as luminous as the Sun; its distance is 85 light-years.

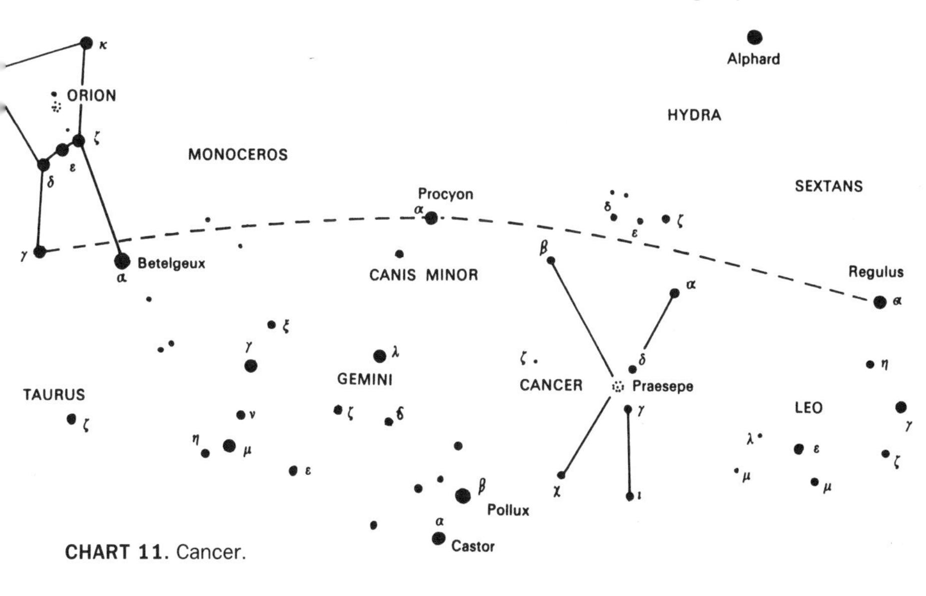

CHART 11. Cancer.

Extending downward from Regulus is a curved line of stars, making up what is called the Sickle (Chart 12, p. 144). Of these the brightest, apart from Regulus itself, is Gamma Leonis or Algieba, which is a fine double. The primary is of magnitude 2.2 and the companion 3.5, and since the separation is over 4 seconds of arc, a small telescope will split the pair. The primary is decidedly orange, with a K-type spectrum. The companion is of type G, and most people describe it as distinctly yellowish, though I admit that I can never see any pronounced colour in it. The pair makes up a binary system, with a relatively long period of 619 years.

The rest of Leo is made up of a triangle of stars, Beta or Denebola (magnitude 2.1), Delta or Zozma (2.6) and Theta or Chort (3.3). Today Denebola is about equal to Algieba, and much fainter than Regulus. If the ancient records are to be trusted, Denebola used to be brighter than it is now, though since it is a normal A-type Main Sequence star I am decidedly sceptical about any real change.

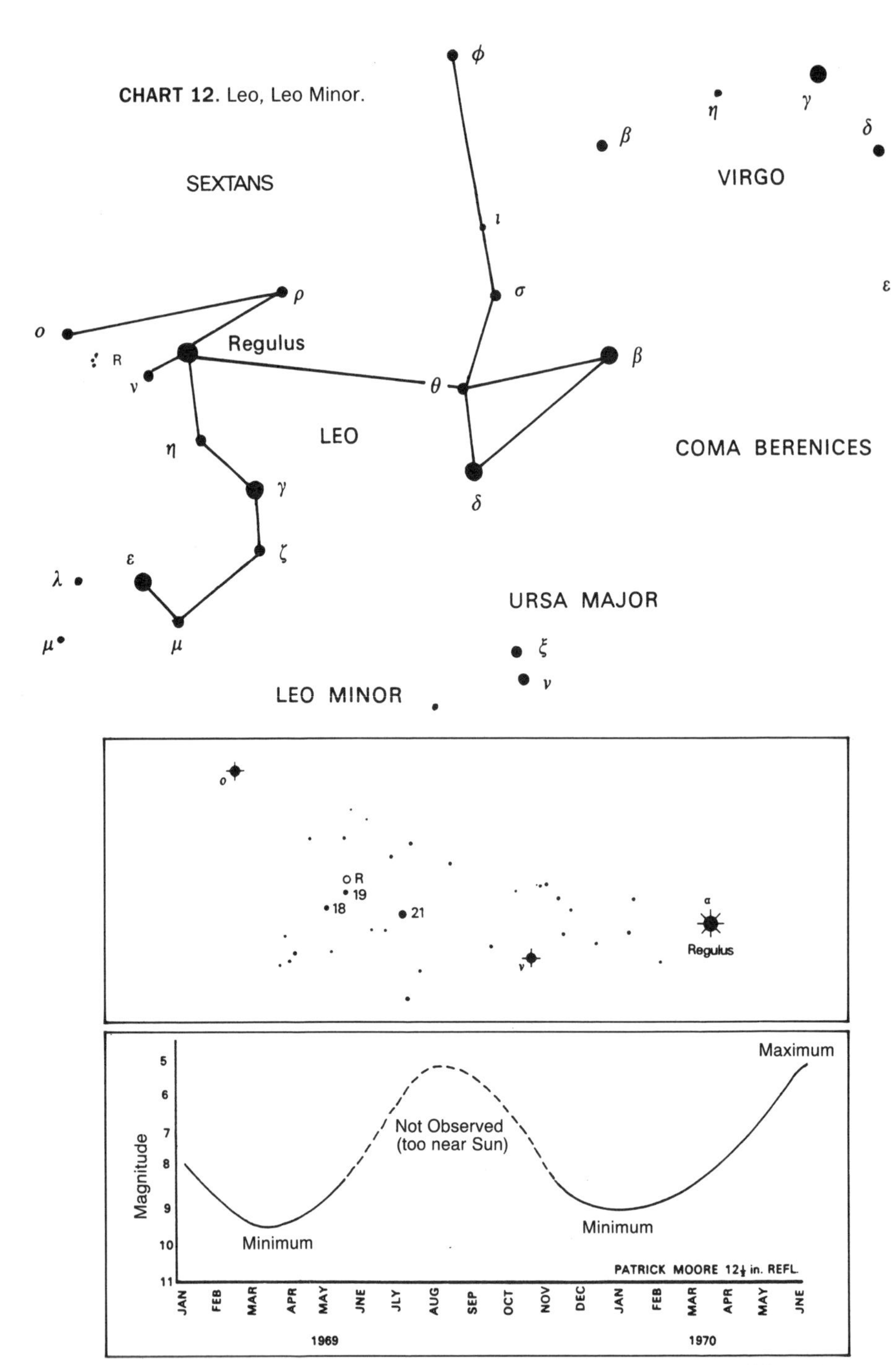

(Above) Chart showing the position of R Leonis. (Below) Light-curve of R Leonis, from observations made in 1969–70 with a 12½-inch (32-cm) reflector.

Before leaving Leo, take out your binoculars, find Regulus, and then swing slightly over to the west, as shown in the diagram. If you are lucky, you will find a group of three faint stars close together, making up a distinctive little group. Of these, the upper member is extremely red. This is R Leonis, a typical Mira-type long-period variable star. It has a range of from magnitude 4.4 to 11.3, and a period of 312 days — though as we have noted, Mira variables are not predictable, and at some maxima R Leonis may never rise to naked-eye visibility. Still, the field is not hard to identify, and you will soon decide whether R Leonis is visible or not. Of course, a relatively small telescope will allow it to be followed throughout its cycle, and there are some convenient comparison stars to hand in 18 Leonis (5.8) and 19 Leonis (6.4). Finally, there are several reasonably bright galaxies in Leo, including four in Messier's catalogue (numbers 65, 66, 95 and 96). Of these the first two are within binocular range, and are in the same wide field with Theta.

The large triangle formed by Regulus, Procyon and Pollux is occupied by the faint Zodiacal constellation of Cancer (the Crab) (Chart 11). The main item of interest here is the open

CHART 13. Hydra (head), Sextans, Monoceros.

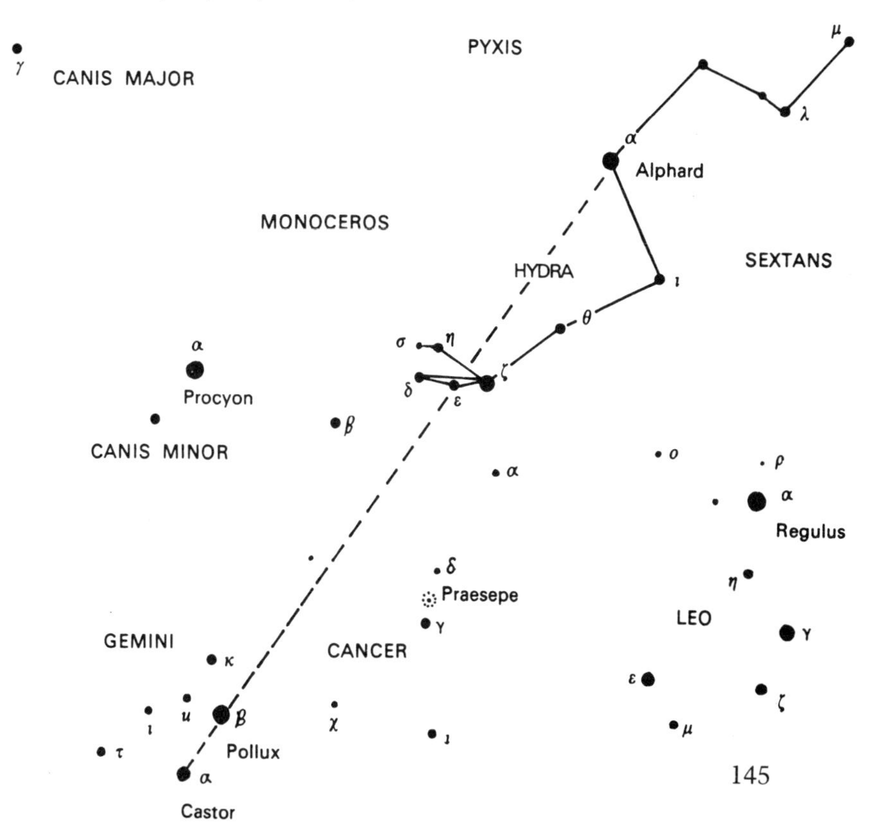

cluster Praesepe, Messier's No.44, which is an easy naked-eye object; binoculars will show it even in full moonlight. Because it has sometimes been nicknamed the Manger, the two stars flanking it, Delta Cancri (magnitude 3.9) and Gamma (4.7), are often called the Aselli, or Asses. Praesepe is a typical loose cluster, richer than the Hyades but not nearly so condensed or spectacular as the Pleiades. Another open cluster in Cancer is M67, near Alpha Cancri or Acubens (magnitude 4.2). M.67 is on the fringe of naked-eye visibility, and seems to be unusually old for a cluster of its type. This is because it is well away from the main plane of the Galaxy, and has been left comparatively undisturbed by passing stars.

In the same binocular field with Delta you will find one of the reddest stars in the sky: X Cancri. It is a semi-regular variable; at maximum it rises to magnitude 5, and it never falls below 7.3, so that it can always be seen with binoculars. It looks rather like a tiny glowing coal!

Stretching overhead is Hydra (the Watersnake) which is immensely long; it begins near Cancer, and ends not far from Centaurus. Now that Argo has been carved up, Hydra has the distinction of being the largest constellation in the sky, but it is also one of the most barren. Its only bright star is the second-magnitude Alphard (Alpha Hydra), nicknamed 'the Solitary One' because it is so isolated. Castor and Pollux point directly toward it, and during autumn evenings it is high up (Chart 13). It is reddish, with a K-type spectrum. Sir John Herschel made some observations of it during his voyage back to England from the

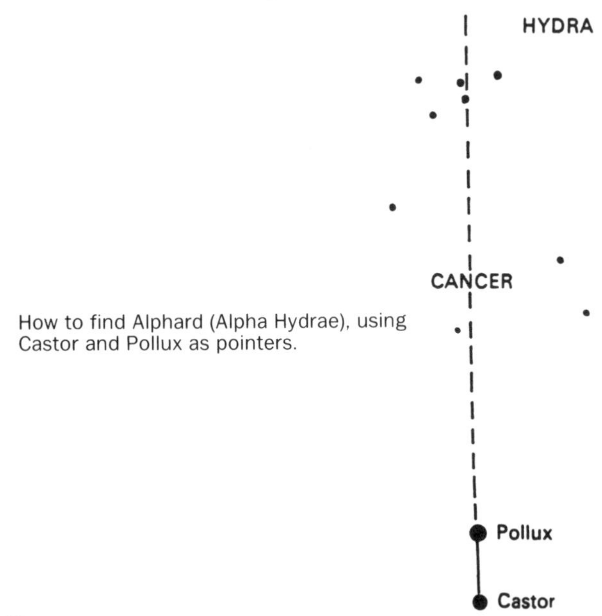

How to find Alphard (Alpha Hydrae), using Castor and Pollux as pointers.

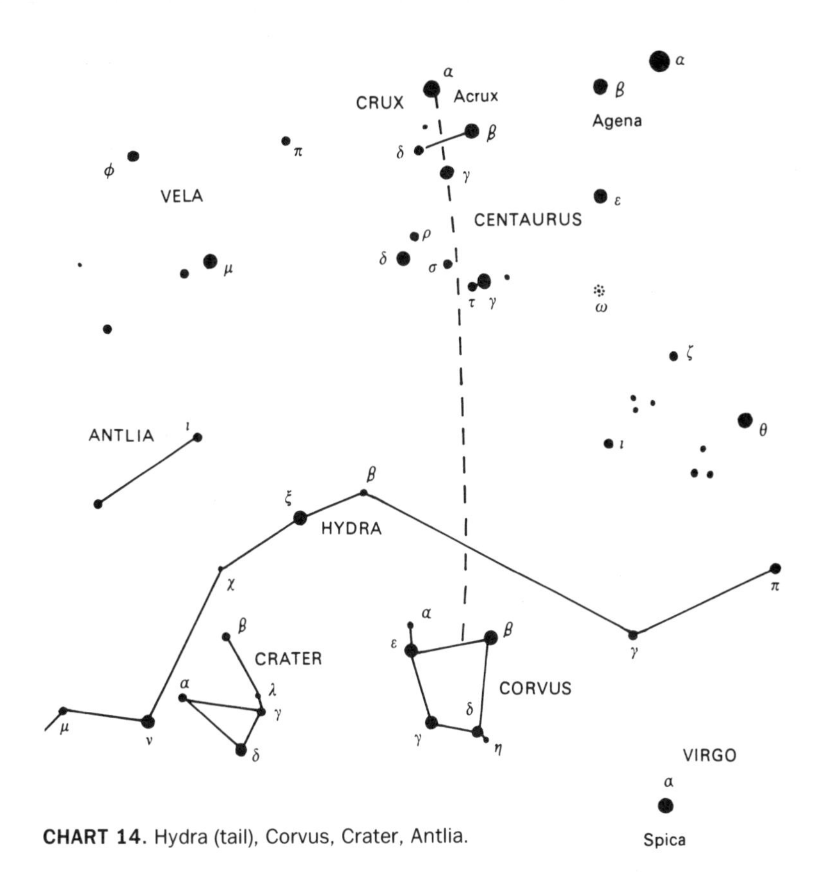

CHART 14. Hydra (tail), Corvus, Crater, Antlia.

Cape, in 1838, and believed it to be variable. This has not been confirmed, but it may be as well to keep a watch, though it is awkward to estimate, because there are no suitable comparison stars anywhere near it. We need spend little time on the rest of Hydra, but the quadrilateral of stars making up Corvus (the Crow), resting on the Watersnake's back, is quite prominent. All four stars (Gamma, Beta, Delta and Epsilon) are between magnitudes 2.5 and 3; Alkhiba, the star lettered Alpha, is below magnitude 4, so that the sequence of Greek letters is well out of order. A good way to identify Corvus is to go back to the Southern Cross and use Alpha and Gamma Crucis as pointers (Chart 14). Adjoining Corvus is Crater (the Cup), whose brightest star (Delta) is only of magnitude 3.6, and which contains no notable objects. Yet another faint, barren group is Sextans, the Sextant, which lies between Leo and Alphard (Chart 13, p. 145). There seems no good reason for Sextans to form a separate constellation. It was introduced to the sky in 1690 by the Danzig observer Hevelius, who apparently constructed sextants and was rather proud of them!

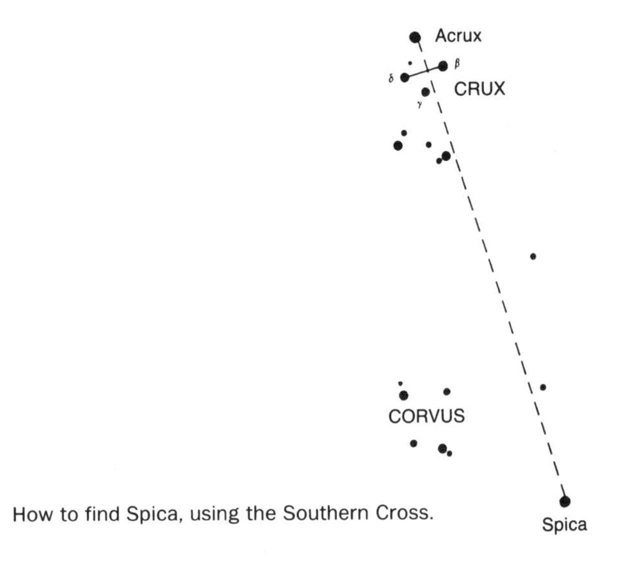

How to find Spica, using the Southern Cross.

In the north-east the brilliant orange Arcturus has come into view, but will be better placed during winter evenings. Higher up is the large and distinctive Zodiacal group of Virgo (the Virgin), led by the white first-magnitude star Spica.

One method of locating Spica is to use the Southern Cross, though the procedure is not quite so straightforward as usual. Begin at Alpha Crucis, and pass a line directly between Beta and Gamma, as shown in the diagram. After crossing a barren region you will arrive at Virgo, and Spica itself is sufficiently bright and solitary to be recognized at once.

Virgo is important enough to deserve a separate map (Chart 15). The main pattern takes the form of an inverted Y, at the base of which is Gamma Virginis or Arich, a fine binary with virtually equal components, each of magnitude 3.5. The present separation is 3 seconds of arc, so that the pair is still easy to split, but it is closing up, not because it is really doing so but because we are observing it from an increasingly unfavourable angle. Not far into the next century the components will seem so close together that they will be difficult to separate at all, though subsequently they will open out again. The orbital period is 171.4 years.

The 'bowl' of the Y, bounded on the far side by Denebola in Leo, is filled with faint galaxies. Virgo contains no less than 11 Messier objects, including the giant elliptical system M.87 and the remarkable M.104, which earns its nickname of 'the Sombrero Hat'. Unfortunately, a large telescope is needed to see any of these galaxies well. Below Virgo is the large, extended cluster marking Coma Berenices (Berenice's Hair), but this is

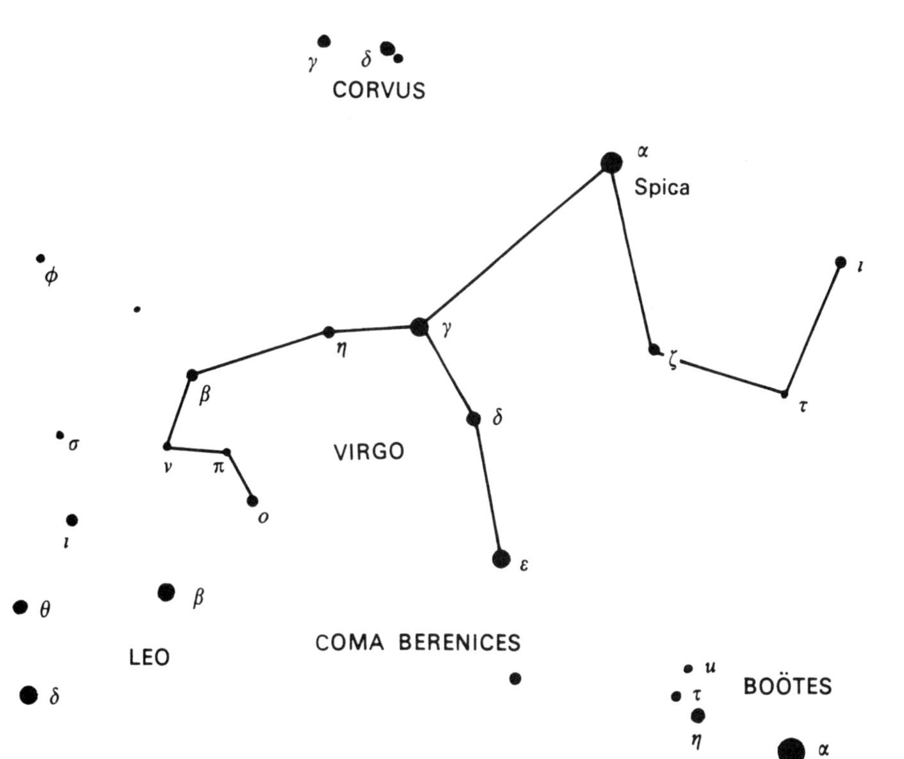

CHART 15. Virgo.

not high enough to be really well seen, at any rate from New Zealand.

Next, let us turn to the southern aspect of the evening sky. The Cross is now high, which means of course that Achernar in Eridanus is low down. The dismembered Argo occupies much of the high south-east, with Canopus gleaming down as though in a serious effort to match the magnificence of Sirius. (Strange to reflect that in fact, Canopus is around 8000 times the more luminous of the two!) Antares in the Scorpion has made its entry in the east, but will be more prominent during winter.

Meantime there is Centaurus (The Centaur), shown in Chart 16 (p. 150). Without doubt it is one of the most splendid constellations in the sky, and it contains a number of bright stars as well as its two leaders. In shape it is distinctive; it seems to straddle the Southern Cross, and does indeed give a vague impression of some large semi-human figure.

I have already said a good deal about Alpha Centauri. To us it is important because it is so close, and moreover it is a superb binary; one component is considerably brighter than the other, and the revolution period is only 80 years, so that both

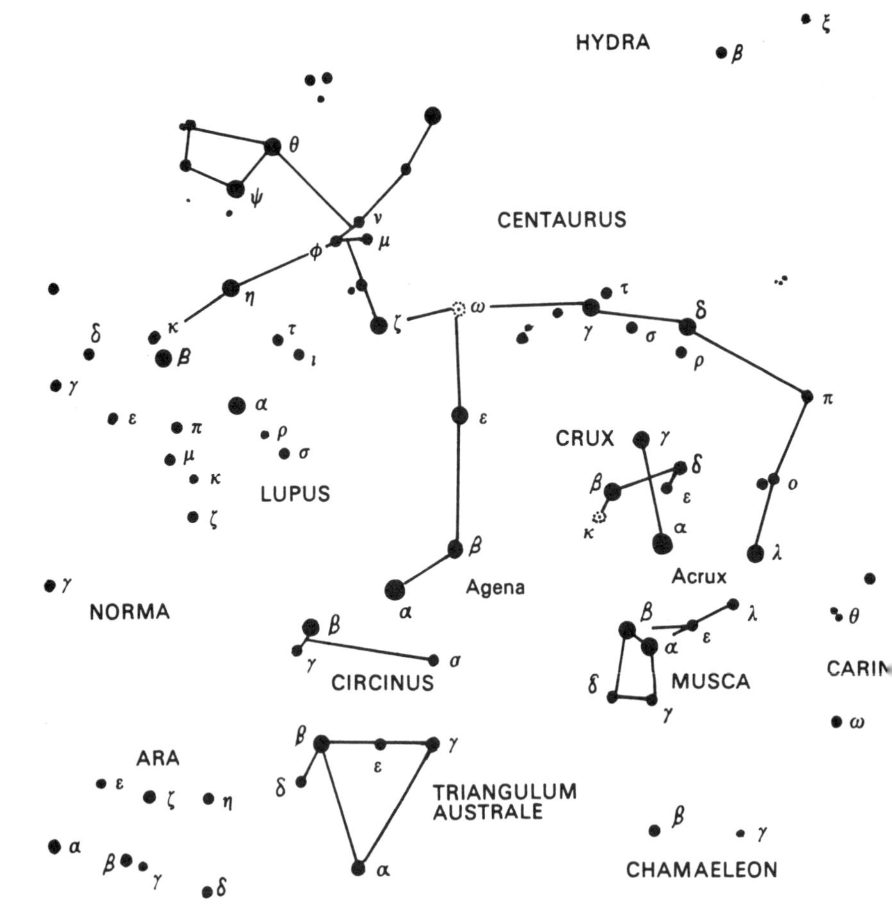

CHART 16. Centaurus, Crux Australis, Triangulum Australe, Musca, Circinus.

the separation and the position angle change quite quickly. Its dim companion, Proxima, is slightly closer to us and, at 4.2 light-years, is actually the closest of all the stars beyond the Sun. It has always been regarded as a distant binary companion of Alpha, but doubts have been expressed recently, and it has been suggested that it is merely a chance neighbour. It is very faint, and none too easy to identify even with a powerful telescope.

Agena or Beta Centauri, the other Pointer, is quite different from Alpha. It is more than 10,000 times as luminous as the Sun, and 460 light-years away — well over a hundred times as remote as Alpha. Between Alpha and Beta is a red Mira-type variable, R Centauri, which has a period of 546 days. At minimum it drops to below the twelfth magnitude, but at maximum it may reach 5.3, in which case binoculars will show it.

Lying in line with Agena and the second-magnitude star Epsilon Centauri is Omega, the brightest globular cluster in the sky and always worth looking at. Near its edge it is easily resolvable into stars, as near its centre the stars are so crowded that they merge into a blur of light. Quite different is the open cluster round Lambda Centauri, not far from the Cross. There is no symmetry here, but the cluster is bright enough to be conspicuous with the naked eye. Finally, make note of Gamma Centauri, which is a binary with equal components (each magnitude 2.9) and an orbital period of 84.5 years. The separation is between 1 and 2 seconds of arc.

Centaurus practically surrounds the Southern Cross, with its unmistakable kite-pattern, its Jewel Box and its Coal Sack. Adjoining the Cross is Musca (the Fly). Here, the brightest star, Alpha Muscae, is only of magnitude 2.7, but there is one pair of stars which merits attention. In the same wide binocular field with Alpha Crucis you will find Lambda and Mu Muscae, each of magnitude 4.7. They have beautifully contrasting colours. Lambda is pure white, while Mu is very red, with an M-type spectrum. Use binoculars to look at them, and you will see what I mean. Of course there is no real connection between them; Mu is more than six times as far away from us as Lambda.

Lupus (the Wolf) contains some reasonably bright stars, of which one (Alpha Lupi) is of magnitude 2.3, but there are not many objects of interest. We need not spend time upon other very obscure groups, such as Circinus (the Compasses), Norma (the Rule) and Telescopium (the Telescope). Ara (the Altar) has a distinctive shape, and of its leading stars two, Beta (2.8) and Zeta (3.1) are decidedly orange.

Few of the constellations bear the slightest resemblance to the object they are meant to represent. It takes a great deal of imagination to make a wolf out of Lupus, a huntsman out of Orion, or a pair of boys out of Gemini. Yet there are exceptions, and one of these is Triangulum Australe (the Southern Triangle) which lies close to Alpha Centauri, and is shown with it in Chart 16. The three main stars, Alpha (magnitude 1.9), Beta (2.8) and Gamma (2.9) do indeed make up a triangle which is definite enough to be more conspicuous than its actual brightness warrants. Alpha is a red giant, 55 light-years away and about 95 times as luminous as the Sun; its colour is detectable with the naked eye and is brought out well in binoculars. It has another distinction, too; with its declination of almost -70° it is the nearest really bright star to the south celestial pole, though it is still much too far away from the polar point to be used as a substitute for the northern hemisphere's obliging Polaris. Over New Zealand and over the southern parts of Australia and South Africa it never sets.

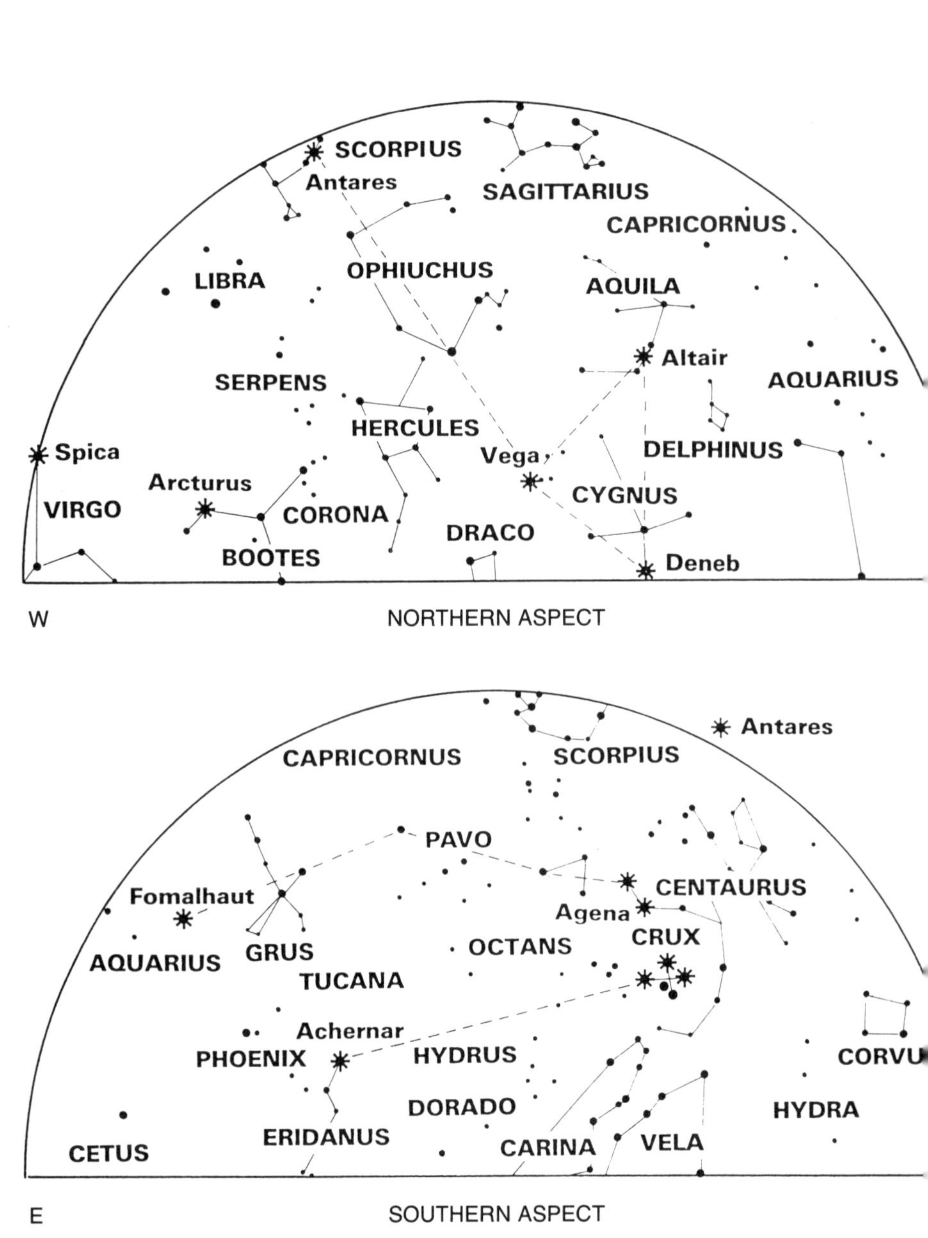

The evening sky in winter.

Chapter 20
Stars Of Winter Evenings

With the onset of winter, we lose some of the best known southern stars from the evening sky, even from New Zealand. Sirius has disappeared below the horizon, and the whole of Orion's retinue is out of sight. The charts given here are for 10 p.m. in mid-July. Anyone who wants to work out the equivalent times for other months can simply apply the 'two-hour' rule mentioned in Chapter 18.

There is a legend that Orion, the mighty hunter, boasted that he could kill any creature on earth. Unfortunately he had forgotten about the scorpion, which crawled from a hole in the ground, stung him in the heel and caused his untimely death. He was rescued by the intervention of the Olympian gods, who placed him in the sky, and with their innate sense of fair play also elevated the scorpion to celestial rank, taking the precaution of putting it as far away from Orion as possible so that there could be no fear of any further unpleasantness! And during winter evenings, with Orion out of view, Scorpius is practically overhead.*

It is a magnificent constellation, made up of a long line of stars which certainly can conjure up the impression of some venomous creature. Its leader, Antares, the 'Rival of Mars', is a huge red supergiant. There are no definite pointers to it, but a guide-line starting from Alpha Crucis and passing through Alpha Centauri will arrive somewhere near it, and this is bound to be good enough. To make Antares even more imposing, it has a fainter star to either side of it, as shown in Chart 17 (p. 155); Tau Scorpii (magnitude 2.8) to the south, Sigma Scorpii or Alniyat (2.9) to the north. Both these are very luminous and remote, and are comparable in power with Antares itself. However, they are much further away, 780 and 600 light-years respectively, as against only 330 light-years for Antares.

This may be the moment to introduce a new term: 'absolute magnitude'. This is the apparent magnitude which a star would have if it could be seen from a standard distance of 32.6 light-years; a distance chosen for reasons which need not concern us for the moment. From this distance Antares would shine as of magnitude -4.7, Tau Scorpii -4.1 and Sigma -4.4, so that they would be strictly comparable with Venus as seen from Earth.

* Many people refer to it as 'Scorpio', but 'Scorpius' is the correct form.

But the absolute magnitude of our Sun is only +4.8, so that it would be a very dim naked-eye object.**

Antares is 7500 times as luminous as the Sun. It has a 5.4-magnitude binary companion at a separation of 2.7 seconds of arc; the real separation is of the order of 75,000 million km. The orbital period is 878 years, and the companion, feeble though it looks, has a full 50 times the power of the Sun. It is often said that the companion star is green, though this is probably due mainly if not entirely to contrast with the fiery hue of Antares itself.

The Scorpion's head is made up of Beta or Graffias (magnitude 2.6), Omega (4.0) and Nu or Jabbah, which is a very wide double; the magnitudes are 4.3 and 6.5, and the separation is over 40 seconds of arc. Graffias has a fifth-magnitude companion at a separation of almost 14 seconds of arc; the main star is itself a very close binary. Zeta Scorpii, much further south, is made up of two, an orange star of magnitude 3.6 and a white star of magnitude 4.7. The fainter member of the pair is much the more luminous, and is a great deal more powerful than Antares, but it is over 2500 light-years away, while the brighter star is within 200 light-years of us. In the same region is another wide pair, Mu Scorpii, separable with the naked eye; the magnitudes are 3.0 and 3.6. As the two components share a common motion through space they are probably physically associated, but they are at least a light-year apart. Another feature of the Scorpion is its 'sting', with two bright stars, Lambda or Shaula (1.6 — only just below the official 'first magnitude') and Upsilon or Lesath (2.7). Again, there is no genuine link, because Lesath is much the more luminous, and is almost six times as far away from us.

The Milky Way is very rich in Scorpius, and there are several bright star-clusters, notably the two globulars M.4 and M.80, not far from Antares, and the lovely open clusters M.6 and M.7, which are naked-eye objects and can be resolved into stars with binoculars; M.6 is nicknamed the 'Butterfly'. Altogether Scorpius is one of the most magnificent constellations in the sky, and some people rank it with Orion. In view of the mythological legend, perhaps this is only fair!

There is a marked contrast with the adjacent constellation of Libra (the Scales or Balance), which is very obscure, and is worthy of note only because it too lies in the Zodiac. The brightest star, Beta Librae or Zubenelchemale, is of magnitude 2.6. It is said to be the only single naked-eye star with a greenish

** To complete the story: 32.6 light-years is equal to 10 parsecs. One parsec (3.26 light-years), is the distance from which a star would subtend a parallax of 1 second of arc. In fact no star, apart from the Sun, is as close as this. The parallax of Alpha Centauri is only 0.76 of a second of arc.

hue, but most people will call it white, and I have never been able to see any colour in it. It has a normal B-type spectrum, and is about 100 times as luminous as the Sun.

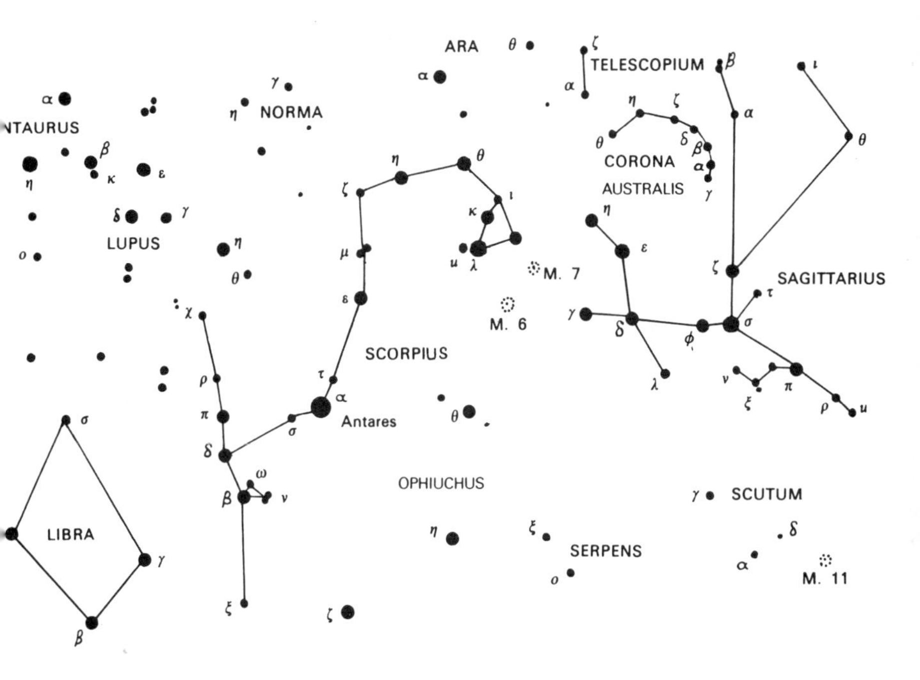

CHART 17. Scorpius, Sagittarius, Libra, Telescopium, Corona Australis.

Immediately east of Scorpius, and also near the overhead point during winter evenings, is Sagittarius, the Archer. It has two particularly bright stars, Epsilon or Kaus Australis (magnitude 1.8) and Sigma or Nunki (2.0), together with several more above the third, but there is no really obvious pattern. It is often nicknamed the Teapot, though I have never been able to understand why! To help locate it, I have included it with Scorpius in Chart 17 instead of giving it a map to itself. Here too is the little semicirclet of stars marking Corona Australis or Corona Austrinus, the Southern Crown. The Crown is near the two stars of Sagittarius lettered Alpha and Beta; both these are surprisingly faint — Alpha (Rukbat) is of magnitude 4.0, and Beta (Arkab) is a wide naked-eye double with components of magnitudes 3.9 and 4.3.

Sagittarius contains more Messier objects than any other constellation, and clusters and nebulae abound, together with the magnificent star-clouds which hide our view of the centre of the Galaxy. There are for example two superb gaseous nebulae, M.20 (the Trifid) and M.8 (the Lagoon), not far from

Lambda Sagittarii, of magnitude 2.8. The Lagoon is an easy binocular object, and a small telescope will show both. They are, moreover, favourite targets for astronomical photographers. Sweep across the Archer with binoculars or a wide-field telescope, and you will see an almost endless succession of superb star-fields.

Then, by contrast, turn toward the adjacent Zodiacal groups of Capricornus (the Sea-Goat) and Aquarius (the Water-bearer), which are large, dim and dull. I have shown them together in Chart 18. The brightest star in Capricornus is Delta (magnitude 2.9). Alpha or Giedi is a naked-eye double, with components of magnitudes 3.6 and 4.2, but are unrelated, as the fainter component is much the more remote. Beta or Dabih has a sixth-magnitude companion at a separation of 205 seconds of arc, so that both components can be seen with binoculars. For once we have an optical pair, not a binary system. Capricornus may be located by using the three stars in the line of which Altair is

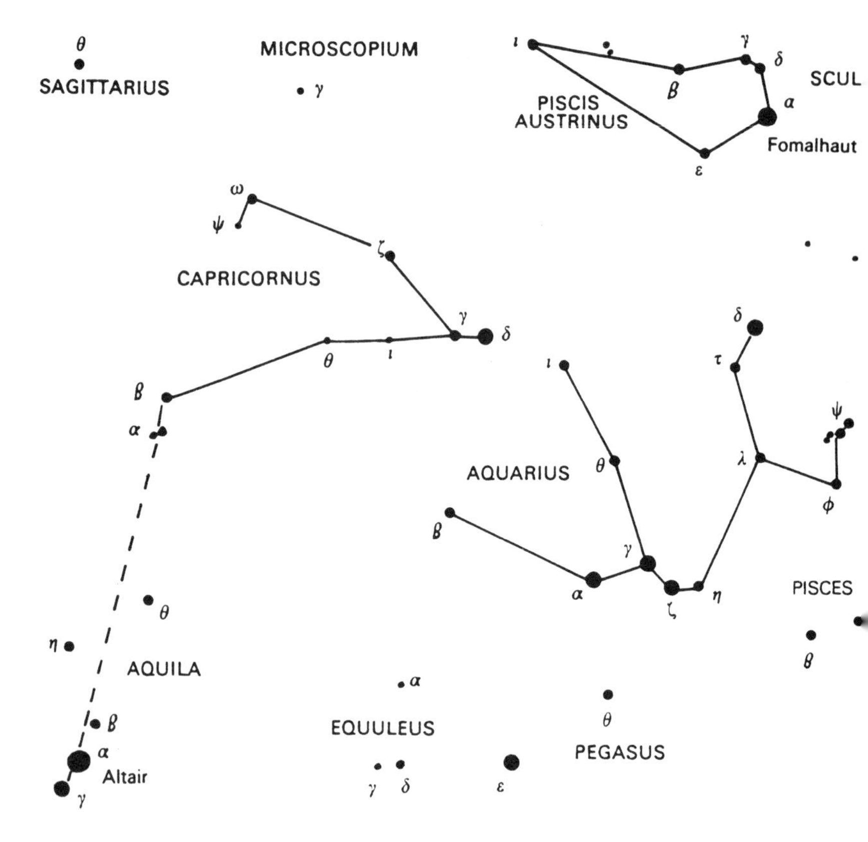

CHART 18. Aquarius, Capricornus, Piscis Australis, Microscopium.

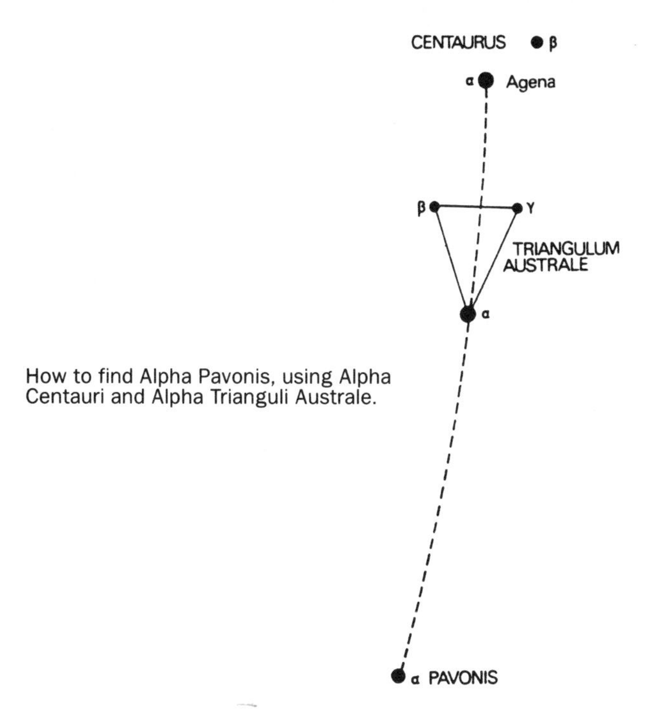

How to find Alpha Pavonis, using Alpha Centauri and Alpha Trianguli Australe.

the central member, as shown in Chart 18. Aquarius is scarcely more distinguished, though not far from the brightest star, Beta Aquarii or Sadalsuud (magnitude 2.9) lies the fine globular cluster M.2, which is only just below naked-eye visibility and is easy to see with binoculars. It is around 55,000 light-years away. It may be worth looking at the little group of stars round Psi Aquarii; several of them are orange, and have occasionally been mistaken for a very loose cluster.

High in the south-east lies Piscis Australis, the Southern Fish (sometimes referred to as Piscis Austrinus). Fomalhaut, the leader, is of the first magnitude. It is 13 times as luminous as the Sun, and at 22 light-years is the nearest of the really bright stars apart from Alpha Centauri, Sirius, Procyon and Altair. It is white and, as we have noted, is surrounded by a cloud of material which may indicate the presence of a planetary system, though it would be dangerous to claim that planets really exist. There are no clear pointers to the Fish, and it contains nothing of interest apart from Fomalhaut itself.

Since we have begun this season's viewing by looking south, let us continue, and come to the Southern Birds, of which there are four: Grus (the Crane), Phoenix (the Phoenix), Tucana (the Toucan) and Pavo (the Peacock). Of all the regions of the sky, this is probably one of the most difficult to sort out, partly because there are no obvious signposts and partly because only Grus has a distinctive shape. The best method, so far as I can work out, is to locate Alpha Pavonis (magnitude 1.9) by using Alpha Centauri and Alpha Trianguli Australe as guides, as shown in the diagram. Continue further and the same line will pass somewhere near Grus, and eventually to Fomalhaut. It is also worth noting that Alpha Pavonis lies roughly midway between Achernar and Sagittarius. All the Birds are shown in Chart 19, together with the entirely obscure Microscopium (the Microscope) and Indus (the Indian), neither of which can be regarded as an inspired addition to an already confusing area of the sky.

There is nothing special about Alpha Pavonis, which is a normal bluish-white star, but there are two objects in the Peacock which are worthy of mention. One is the globular

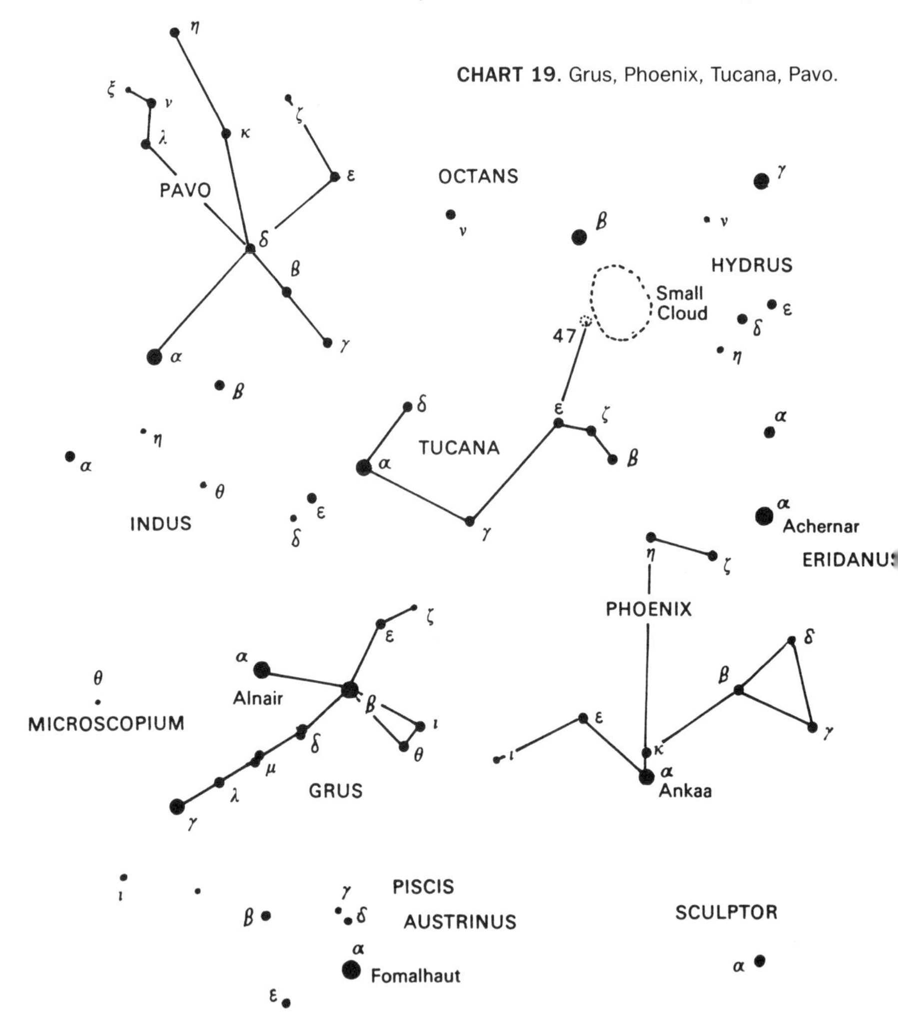

CHART 19. Grus, Phoenix, Tucana, Pavo.

cluster NGC 6752, not far from Lambda. The other is Kappa Pavonis, which is a type II Cepheid variable and is therefore much less luminous than a 'classical Cepheid' would be if it had the same period of 9.1 days. The range is from magnitude 3.9 to 4.7; convenient comparison stars are Delta (3.6), Epsilon (4.0) and Zeta (4.1). In the adjacent Bird, Tucana, we find most of the Small Cloud of Magellan — and, remember, it was by studying the short-period variables in the Cloud that Henrietta Leavitt was able to find the vitally important link between the period and the real luminosity of a Cepheid.

Almost silhouetted against the Cloud are two globular clusters, NGC 104 (otherwise and more generally referred to as 47 Tucanae) and NGC 362. 47 Tucanae is the finest of all the globulars apart from Omega Centauri, and telescopically I find it more beautiful than Omega, because it does not completely fill an average telescopic field. It is rich in variable stars and also contains a surprising number of very rapid pulsars! Use binoculars and you will see that its surface brightness is much higher than that of the Cloud. NGC 362 is also very fine, though below naked-eye visibility.

Tucana contains one very wide, easy double star, Beta, in which the components are more or less equal at magnitude 4.5, and can be seen separately without optical aid. The brighter member is itself double, and the fainter component is a very close binary, so that Beta Tucanae is quite a complicated group. The leader of Phoenix is Alpha, or Ankaa, of magnitude 2.4. The only interesting object is the Algol eclipsing binary Zeta, with a range of from magnitude 3.6 to 4.4 and a period of 1.7 days. Suitable comparison stars are Eta Phoenicis (4.4), Epsilon (3.9) and Beta (3.3). Ankaa lies slightly off the mid-point of a line joining Achernar to Fomalhaut, which is a fairly good way of identifying it. We can pass over Sculptor (the Sculptor) which adjoins Phoenix and is entirely unremarkable, but Grus, the celestial Crane, is much more imposing and gives a superficial impression of a bird in flight. During winter evenings it is high up, somewhat east of south.

The two chief stars in Grus are Alpha or Alnair (magnitude 1.7) and Beta or Al Dhanab (2.1). They are very dissimilar. Alnair is bluish-white, 68 light-years away and 230 times as luminous as the Sun; Al Dhanab can match 750 Suns, but is 173 light-years from us, so that it is much further away from Alnair than we are. It is an orange-red giant, and with binoculars it is interesting to swing from it to Alnair. The colour contrast is even greater than that between Alpha and Gamma Crucis.

Forming a rough triangle with Alnair and Al Dhanab is a very wide double, Delta Gruis. The two are easy to split with the naked eye. They are not related — the fainter and closer member of the pair is red, while the brighter star is a yellowish giant. Further along the 'line' of Grus is another pair of the same type, Mu Gruis. The Crane is rich in galaxies, but all these are below the tenth magnitude.

I hope that you will have no trouble in sorting out these Birds. Once they have been studied carefully, they are identifiable enough. Before leaving them, it is worth mentioning that Achernar, Ankaa, Alnair and Alpha Pavonis form a large quadrilateral. Once this has been located, everything else will fall neatly into place.

There is no need to say more at present about the Centaur and the Cross, which are in the south-west. Instead, let us turn to the northern aspect. The famous Square of Pegasus is coming into view, but it is really a spring group. Low in the north, observers in parts of Australia and South Africa may be able to make out part of Draco (the Dragon), but even at its highest, its brightest and most northerly star, Gamma (Eltamin), never reaches more than a few degrees above the horizon, so it is completely out of view from New Zealand. However, we must pause to look at the constellation of Boötes, the Herdsman (Chart 20) which is headed by Arcturus and is well above the north-western horizon.

Arcturus is extremely brilliant. In fact it is surpassed only by Sirius, Canopus and Alpha Centauri; its magnitude is minus 0.04, and it is a lovely light orange colour. It is 36 light-years away and 116 times as luminous as the Sun. The rest of Boötes is not of special interest, though Epsilon Boötis or Izar

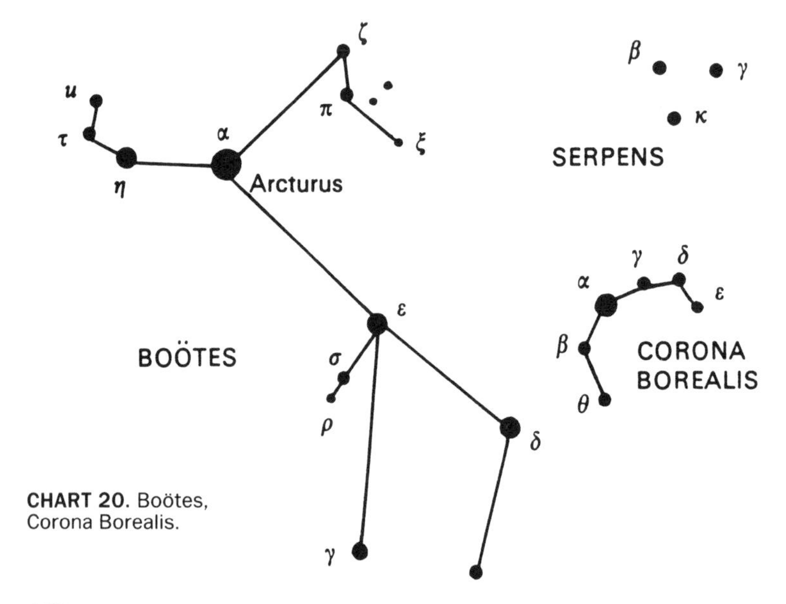

CHART 20. Boötes, Corona Borealis.

(magnitude 2.4) is much the same colour as Arcturus, and has a fifth-magnitude companion at a separation of 2.8 seconds of arc. Adjoining Boötes is Corona Borealis (the Northern Crown), which has one second-magnitude star, Alpha (Alphekka) and consists of a prominent little semi-circle, much more definite and considerably brighter than the Southern Crown near Sagittarius. It also contains two important variables which have already been described; the 'sooty star' R Coronae, in the bowl of the Crown, and the Blaze Star, T Coronae, which is one of the rare recurrent novae.

It is possible to locate Arcturus by using Alpha and Beta Crucis as pointers, but the line is very long and somewhat curved, because the Herdsman and the Cross are on opposite sides of the sky. Yet once Arcturus has been found, it will not be forgotten, if only because of its brilliance. Incidentally, it is sufficiently near the celestial equator to be seen from every permanently inhabited continent. Only from parts of Antarctica does it remain below the horizon.

To the north-east we come to three really interesting constellations: Cygnus (the Swan), Lyra (the Lyre or Harp) and Aquila (the Eagle). I have shown them all on Chart 21 (p. 162), together with the various smaller groups in the same area.

Each of the three large constellations includes a first-magnitude star. Lyra has as its leader the brilliant Vega (magnitude 0.0); Aquila has Altair (0.8) and Cygnus has Deneb (1.2). In Britain these three make up what is known unofficially as the Summer Triangle. This was a term which I introduced casually in a television broadcast long ago, and which seems to have come into general use, though obviously it is parochial. (Remember, June is summertime in England!)

Of the three, only Altair is close enough to the equator to attain a respectable altitude. Vega and Deneb are much further north, and though they can be seen from Australia, South Africa and most of New Zealand, they do not rise from Invercargill.

Beta and Alpha Centauri give a rough guide-line to Altair, but there is a great deal of sky to be traversed, and several notable constellations, such as Sagittarius, lie in between. However, Altair has one characteristic which allows for quick identification. Like Antares, it has a fainter star on either side of it; in this case Gamma Aquilae or Tarazed (magnitude 2.7) and Beta or Alshain (3.7). There can be no possible confusion with Antares, because Altair is pure white. It is also one of our nearest stellar neighbours, at 16.6 light-years, and it is only 10 times as luminous as the Sun. Of its two 'flanking' stars, Tarazed is obviously orange in colour.

Extending from Altair, the rest of the Eagle spreads out in a somewhat birdlike fashion. There are three stars lying in a

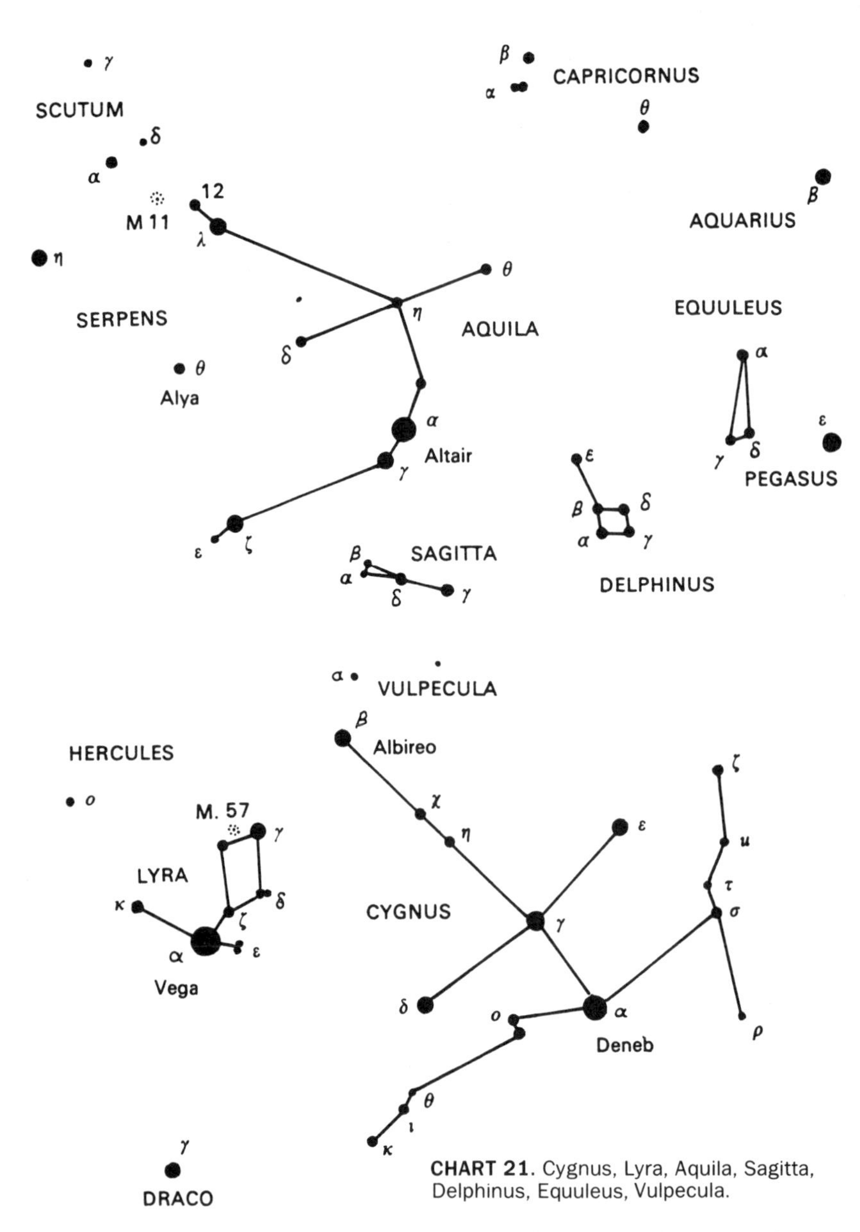

CHART 21. Cygnus, Lyra, Aquila, Sagitta, Delphinus, Equuleus, Vulpecula.

straight line; Theta (magnitude 3.2), Delta (3.4) and Eta Aquilae, the central member, which is a classical Cepheid variable with a range of from 3.6 to 4.4, and a period of 7.2 days.

At the far end of Aquila, near Lambda (magnitude 3.4) is the little constellation of Scutum, the Shield. It has no bright star, but it does contain a glorious open cluster, M.11, nicknamed the 'Wild Duck'. Actually it is somewhat fan-shaped and contains hundreds of faint stars. It can be seen with the naked eye, though it is not too easy to identify because it lies in a rich part of the Milky Way, but telescopically it is a glorious sight. It is well over 5000 light-years away. Also in the Shield is an interesting variable star, R Scuti, which has alternate deep and shallow minima interspersed with spells of total irregularity. The mean period is 140 days, and the range from magnitude 5 to 6, but every fourth or fifth maximum is lower, and the star may drop to the eighth magnitude. It is what is termed an RV Tauri variable, after the prototype member of the class; it is a true variable, not an eclipsing binary. RV Tauri stars are rare, and R Scuti is much the brightest of them. It seems to be oscillating in at least two superimposed periods, and at its peak it must be at least 8000 times as luminous as the Sun.

Vega, in Lyra, is much brighter than Altair, but it is also much further north, and you will not see it from any of New Zealand's South Island. This is a pity because it is a lovely star, steely-blue in colour; as we have noted, it is also one of those stars associated with possibly planet-forming material. Beside it is the double-double or quadruple star Epsilon Lyrae. Also in the constellation is Sheliak or Beta Lyrae, an eclipsing binary which is quite different in type from Algol, because the components are almost touching and are not very unequal. This leads to alternate deep and shallow minima, but there is no similarity with the intrinsically variable R Scuti. Between Sheliak and its third-magnitude neighbour Gamma Lyrae is M.57, the Ring Nebula. This is the best-known of all planetary nebulae, but its low altitude makes it a difficult object from anywhere in Australia and it is, of course, invisible from most of New Zealand.

Deneb, leader of the Swan, is not as imposing as either Vega or Altair, but this is purely because it is so much more remote. It is a particularly powerful star, around 70,000 times as luminous as the Sun, and is 1800 light-years away, so that we now see it as it used to be at the time when Britain was occupied by the Romans. It has an A-type spectrum, and is white. Near it is NGC 7000, the so-called North America Nebula. When photographed with adequate equipment it really does recall the outline of the North American continent. It is dimly visible with the naked eye in the guise of a slightly brighter section of the Milky Way, and binoculars show it clearly as a large region of diffuse nebulosity. The nebula is nearly 50 light-years in diameter, and may owe much of its illumination to Deneb.

Cygnus is often called the Northern Cross, and it really does look more like an X than its southern namesake. The faintest of the five stars in the X, Albireo or Beta Cygni, is also the most noteworthy. It is a glorious, wide double, separable with good binoculars. The brighter component is golden yellow, while the companion is vivid blue. The contrasting colours make it probably the most striking pair in the whole of the sky, and it is not hard to locate. It is of magnitude 3.1, and lies not far away from a line joining Vega to Altair. Between it and Sadr or Gamma Cygni, the central star of the X (magnitude 2.2) lies Chi Cygni, a Mira variable with a period of 407 days. It has an exceptionally large range. At some maxima it has reached almost the third magnitude, while at others it barely exceeds the fifth. At minimum it falls to magnitude 14, and is beyond the range of a small telescope. It is visible with the naked eye for only a few weeks in each year, and its low altitude makes it awkward to find even when at its best. It is, incidentally, an exceptionally strong infra-red source. Eta Cygni (magnitude 3.9) makes a useful comparison star.

The whole of Cygnus is very rich, as the Milky Way flows right through it. There are also some dark rifts, due to unilluminated nebulae, though there is nothing to rival the Coal Sack in the Southern Cross.

There are various small constellations in the area. Delphinus (the Dolphin) is so compact that it looks almost like a star-cluster; Sagitta (the Arrow) has a fairly distinctive shape; Vulpecula (the Fox) contains M.27, the Dumbbell, which is a splendid example of a planetary nebula, and really does recall the shape of the object after which it is named. A modest telescope will show it. Equuleus (the Foal) is dim and undistinguished.

If you draw an imaginary triangle connecting Vega, Arcturus and Antares (Chart 22) you will not trap many bright stars inside. The area is occupied by three constellations which are large but faint: Hercules, Ophiuchus (the Serpent-bearer) and Serpens (the Serpent), all of which are now in the northern sky and extend almost to the overhead point. The only second magnitude star is Ras Alhague or Alpha Ophiuchi (2.1), which lies more or less in the middle of our triangle. Incidentally, Ophiuchus crosses the Zodiac between Scorpius and Sagittarius, so that planets can sometimes pass through it.

There are not many interesting objects in the area. Alpha Herculis (Ras Alhague), close to Alpha Ophiuchi, is a red supergiant which fluctuates between magnitudes 3 and 4. It has a small companion which looks greenish by contrast and is not a difficult object to identify because the separation is almost 5 seconds of arc. Lower down is the globular cluster M.13, which can just be glimpsed with the naked eye under good conditions, but it is by no means the equal of Omega Centauri or 47 Tucanae. Another bright globular, M.5, lies in Serpens; not too far from Alpha Serpentis or Unukalhai (magnitude 2.6).

Also in the Serpent is Alya or Theta Serpentis, a wide, very easy double. The components are perfect twins, at magnitude 4.5, and the separation is over 22 seconds of arc, making it one of the most spectacular pairs in the sky even though both components are white. Alya is more or less in line with the three stars in Aquila above Altair (Delta, Eta and Theta Aquilae). To help identification, I have shown Alya on Chart 21 (p. 162) as well as on its own map in Chart 22.

Note that there are two parts of Serpens: Caput (the Head) and Cauda (the Body). They are separated by Ophiuchus, the Serpent-bearer. It seems that Ophiuchus and Serpens have been engaged in a life-and-death struggle, and that the luckless reptile has come off worst during the encounter!

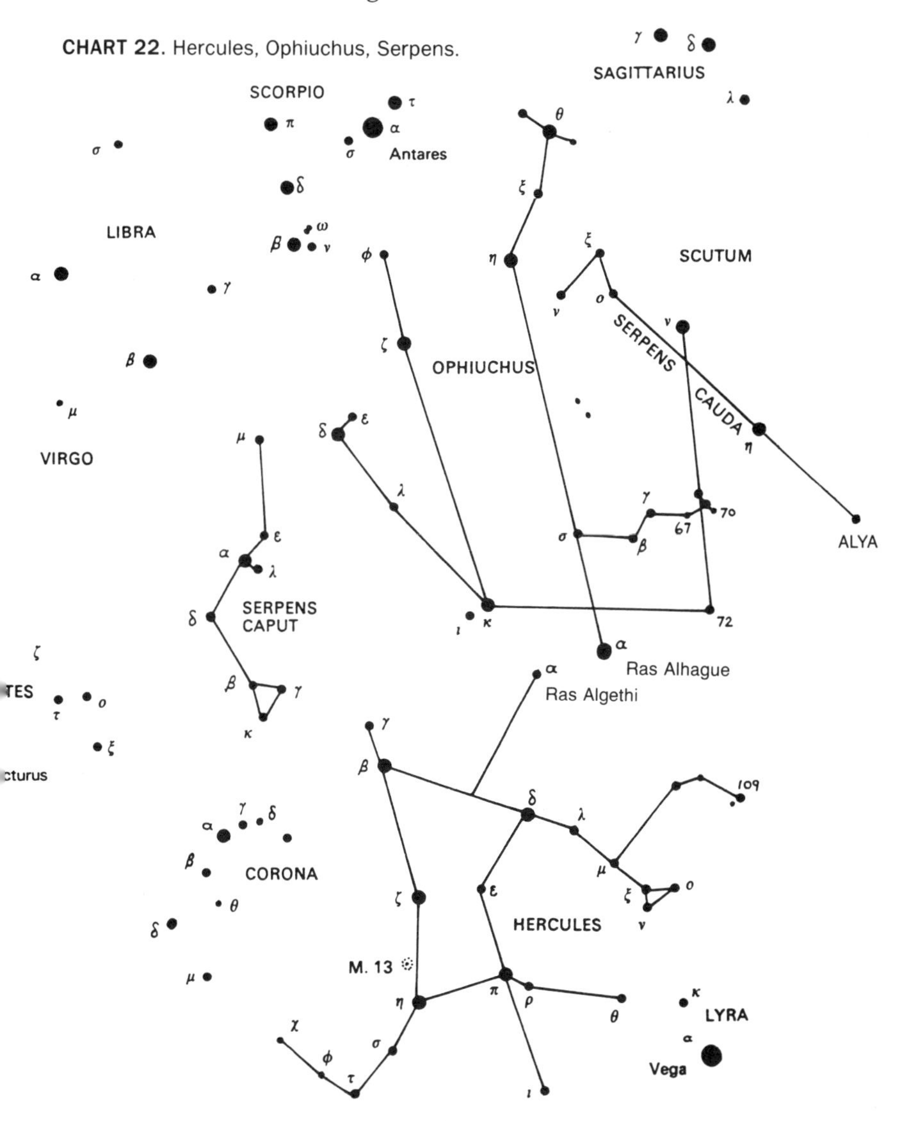

CHART 22. Hercules, Ophiuchus, Serpens.

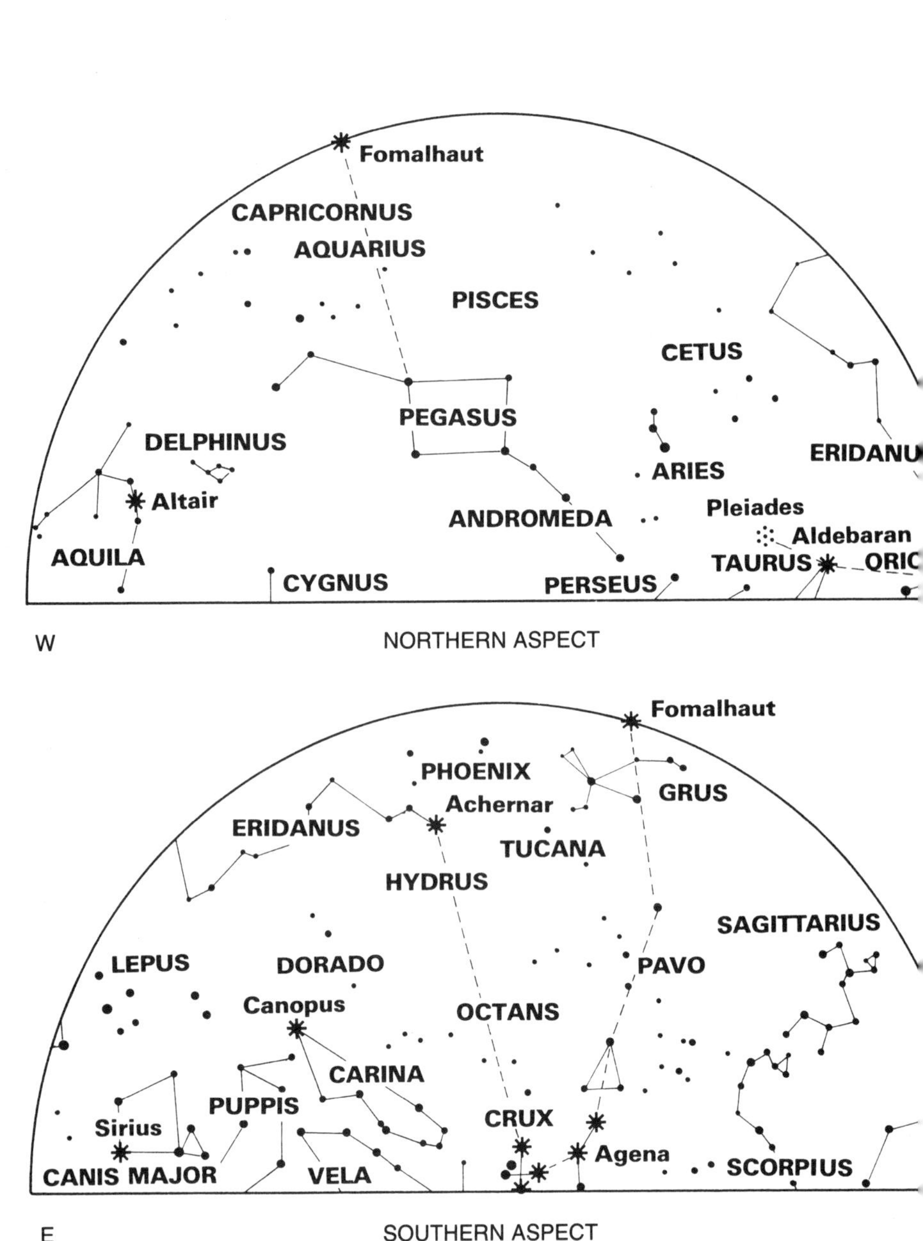

The evening sky in spring.

Chapter 21
Stars Of Spring Evenings

Finally in our 'review of the year' we come to spring. The maps given here are for 10 p.m. in mid-October; they are also valid for midnight in mid-September, 8 p.m. in mid-November, and so on.

The first thing to note is that the Southern Cross is at its lowest, and from latitudes north of Pretoria or Brisbane it actually sets, though admittedly it does not remain below the horizon for very long before it rises again. Centaurus, too, is low, and somehow the picture seems incomplete, as though a vital link is missing. Even Canopus is badly placed, though it has started to gain altitude. Scorpius is disappearing in the south-west; the Southern Birds are prominent, with Phoenix not too far from the zenith, and both Achernar and Fomalhaut are high. At least Orion is returning, and is well on view by midnight.

In the northern part of the sky we have one particularly notable group: Pegasus, the mythological Flying Horse, whose four chief stars make up the celebrated Square. To be precise, one of the four, Alpheratz, has been transferred to the adjacent constellation of Andromeda and is now known as Alpha Andromedae rather than Delta Pegasi, which seems illogical, since it is so clearly a part of the Pegasus pattern. It is of magnitude 2.1. The other stars of the Square are Alpha Pegasi or Markab (2.50), Gamma or Algenib (2.8) and Beta or Scheat, an orange giant which varies rather erratically between magnitudes 2.3 and 2.8 (there is a very rough period of just over five weeks). The only other bright star in Pegasus is Epsilon or Enif (2.4), which lies some way away from the Square.

The Square itself is conspicuous enough, though perhaps not quite so striking as it looks on the map (Chart 23, p. 168). Look inside it and see how many stars you can count with the naked eye. Then use binoculars and see how many more come into view. The difference is quite remarkable.

Scheat and Markab point to Fomalhaut, in Piscis Australis. This is well shown in Chart 24 (p. 169), but is not clear in the hemispherical maps, because Fomalhaut lies in the southern aspect and Pegasus in the northern. Alpheratz and Algenib point to Diphda or Beta Ceti in the Whale (also Chart 24), though here the alignment is not so exact. Cetus is one of the largest constellations in the sky, but certainly not one of the richest.

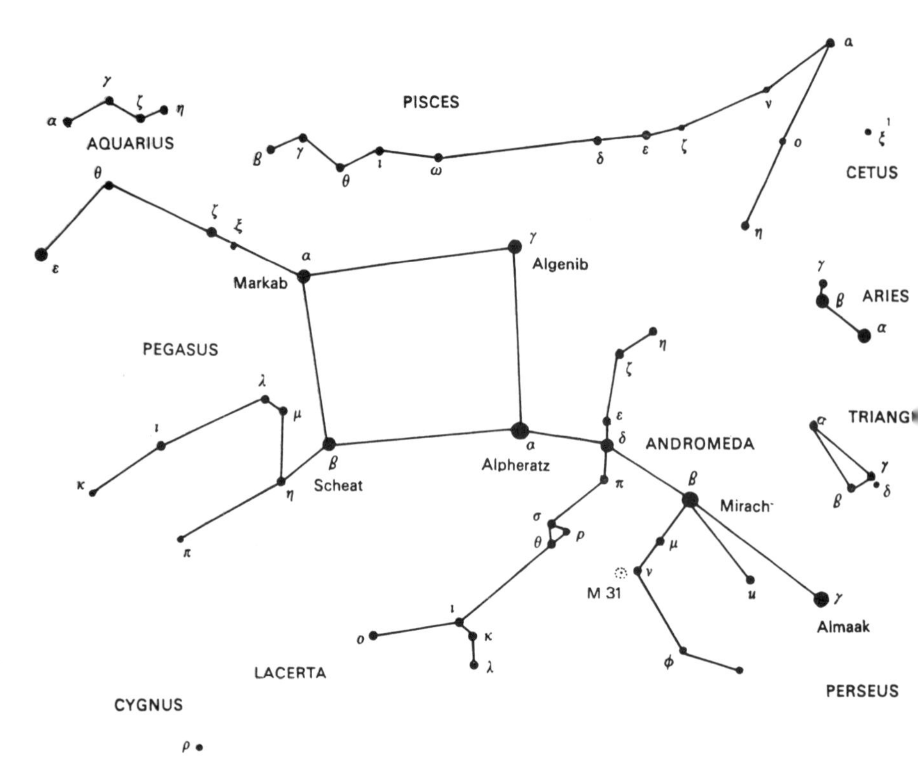

CHART 23. Pegasus, Andromeda, Pisces.

As well as Diphda (magnitude 2.0) it has only one other star above the third magnitude, the reddish Alpha or Menkar (2.5). Tau Ceti (3.5) is a close neighbour, and so is one of the 'Ozma' stars (the other was Epsilon Eridani). Of course, the most celebrated object in the Whale is Mira (Omicron Ceti), the prototype long-period variable which I have already discussed. When it reaches the second magnitude, as it occasionally does, it changes the whole appearance of that part of the sky.

Pisces, the Fishes, lies between Cetus and Pegasus. It is marked by a long line of faint stars, but its sole claim to fame is that it lies in the Zodiac. Fornax, the Furnace, between Cetus and Eridanus, is even less notable, but two of the smaller groups in the area are more interesting. Aries (the Ram) has one second-magnitude star, Alpha Arietis or Hamal (2.0); Gamma, or Mesartim, is a particularly wide and easy double, with components which are equal at magnitude 4.8. The separation is almost 8 seconds of arc. Triangulum (the Triangle) merits its name. The brightest star, Beta Trianguli, is only of magnitude 3.0, but the constellation contains the spiral galaxy M.33, which is a member of the Local Group. It has low surface brightness,

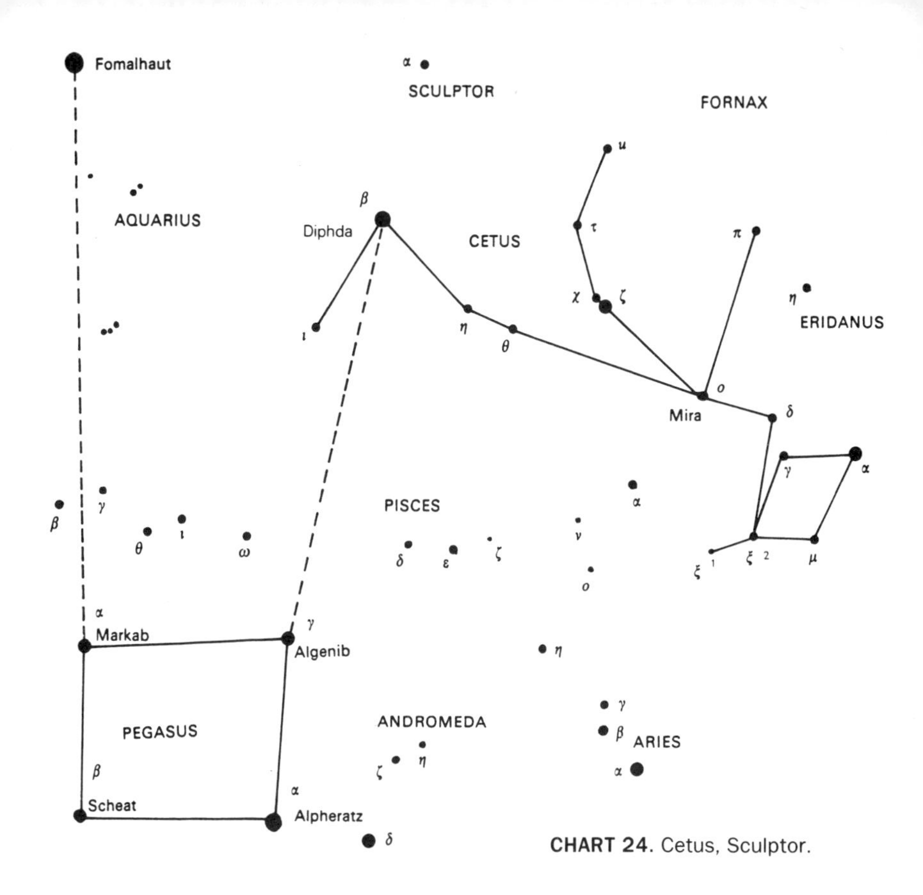

CHART 24. Cetus, Sculptor.

and can be an awkward object to find with a telescope. In fact, it is probably easier to locate with binoculars!

Andromeda, the princess of the ancient legend, extends from Pegasus in the direction of Perseus. Apart from Alpheratz there are two other bright stars, Beta or Mirach and Gamma or Almaak, both of magnitude 2.1. Mirach is orange, while Almaak is orange-red and has a fifth-magnitude companion at a separation of over 9 seconds of arc. The fainter member of the pair is itself a close binary, but is not at all easy to split.

Andromeda is well north of the celestial equator and is very awkward to observe, at least from New Zealand. It is not particularly rich, but it does include one of the most famous objects in the sky: M.31, the Great Spiral, which can be seen (just!) with the naked eye but is more easily located with the help of binoculars. I have given a larger-scale, more detailed map of the area because almost everyone wants to identify the Spiral, even though it is the reverse of spectacular. Locate Mirach and then swing your binoculars down to Mu Andromedae and then to Nu. You should find the Spiral without much trouble, but

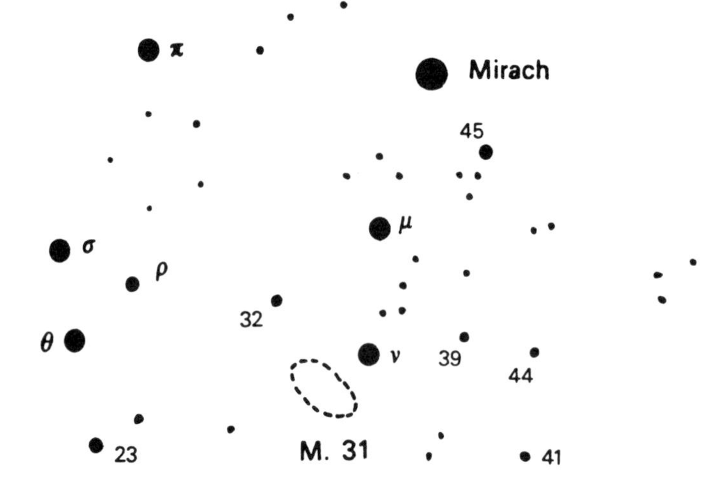

Position of M.31, the Great Spiral in Andromeda.

do not expect it to look anything but a vague blur. At its distance of 2.2 million light-years it is the most remote object to be clearly visible with the naked eye — and when you observe it, you are looking back more than two million years in time.

I hope that you have enjoyed this tour of the sky, and that it has persuaded you to do more than take a casual upward glance on a clear night. Astronomy is surely the best of all hobbies, and if you decide to take a real interest, I am sure that you will not regret it.

I wish you many, many happy hours among the stars.

Appendices

Appendix 1. Planetary Data

Planet Mean distance from Sun, millions of km. Orbital period. Rotation period (equatorial). Orbital Eccentricity.

Planet	*Mean distance from Sun, millions of km*	*Orbital period*	*Rotation period (equatorial)*	*Orbital eccentricity*
Mercury	57.9	87.97 days	58.646 days	0.206
Venus	108.2	224.7 days	243.16 days	0.007
Earth	149.6	365.3 days	23h 56m 04s	0.017
Mars	227.9	687.0 days	24h 37m 23s	0.093
Jupiter	778	11.86 years	9h 50m 30s	0.048
Saturn	1427	29.46 years	10h 13m 59s	0.056
Uranus	2870	84.01 years	17h 14m	0.047
Neptune	4497	164.8 years	16h 7m	0.009
Pluto	5900	247.7 years	6 days 9h 17s	0.248

	Orbital inclination, degrees	*Axial inclination, degrees*	*Escape velocity, km/s*	*Mass, Earth = 1*
Mercury	7.0	2	4.25	0.055
Venus	3.4	178	10.36	0.815
Earth	0	23.44	11.18	1
Mars	1.8	24.0	5.03	0.11
Jupiter	1.3	3.1	60.22	317.9
Saturn	2.5	26.7	32.26	95.2
Uranus	0.8	98	22.5	14.6
Neptune	1.8	28.8	23.9	17.2
Pluto	17.15	122.5	1.18	0.002

	Volume, Earth = 1	*Density, water = 1*	*Surface gravity, Earth = 1*	*Surface temperature, °C*
Mercury	0.056	5.44	0.38	+427
Venus	0.86	5.25	0.90	+480
Earth	1	5.52	1	+22
Mars	0.15	3.94	0.38	-23
Jupiter	1319	1.33	2.64	-150
Saturn	744	0.71	1.16	-180
Uranus	67	1.27	1.17	-214
Neptune	57	2.06	1.2	-220
Pluto	0.01	2.03	0.06	-230

	Diameter, km (equatorial)	*Number of satellites*	*Max.mag.*
Mercury	4878	0	-1.9
Venus	12,104	0	-4.4
Earth	12,756	1	—
Mars	6794	2	-2.8
Jupiter	143,884	16	-2.6
Saturn	120,536	18	-0.3
Uranus	51,118	15	+5.6
Neptune	50,538	8	+7.7
Pluto	2324	1	+14

Appendix 2. Satellite Data

	Satellite Mean distance from primary, km	*Orbital period, d * = retrograde*	*Diameter (mean), km*	*Mean opposition magnitude*
Earth				
Moon	384,400	27.321661	3476	-12.7
Mars				
Phobos	9270	0.3189	22	11.6
Deimos	23,400	1.2624	12	12.8
Jupiter				
Metis	127,960	0.295	40	17.4
Adrastea	128,980	0.298	21	18.8
Amalthea	181,300	0.498	184	14.1
Thebe	221,900	0.675	100	15.5
Io	421,600	1.769	3642	5.0
Europa	670,900	3.551	3130	5.3
Ganymede	1,070,000	7.155	5268	4.6
Callisto	1,880,000	16.689	4806	5.6
Leda	11,094,000	238.7	10	20.2
Himalia	11,480,000	250.6	170	14.8
Lysithea	11,720,000	259.2	24	18.4
Elara	11,737,000	259.7	80	16.7
Ananke	21,200,000	631*	20	18.9
Carme	22,600,000	692*	30	18.0
Pasiphae	23,500,000	735*	36	17.7
Sinope	23,700,000	153*	28	18.3
Saturn				
Pan	133,600	0.57	20	—
Atlas	137,670	0.602	33	18.1
Prometheus	139,350	0.613	105	16.5
Pandora	141,700	0.629	87	16.3
Janus	151,470	0.695	119	14.5
Epimetheus	151,420	0.694	179	15.5
Mimas	185,540	0.942	400	12.9
Enceladus	238,040	1.370	498	11.8
Tethys	294,670	1.888	1046	10.3
Telesto	294,670	1.888	23	19.0
Calypso	294,670	1.888	21	18.5
Dione	377,420	2.737	1120	10.4
Helene	377,420	2.737	35	18.5
Rhea	527,040	4.518	1528	9.7
Titan	1,221,860	15.945	5150	8.4
Hyperion	1,481,100	21.277	288	14.2
Iapetus	3,561,300	79.331	1436	10-12
Phoebe	12,954,000	550.4*	220	16.5
Uranus				
Cordelia	49,471	0.330	26	
Ophelia	53,796	0.372	30	
Bianca	59,173	0.433	42	
Cressida	61,777	0.463	62	
Desdemona	62,676	0.475	54	
Juliet	64,352	0.493	84	
Portia	66,085	0.513	108	

	Satellite Mean distance from primary, km	*Orbital period,* d * = *retrograde*	*Diameter (mean), km*	*Mean opposition magnitude*
Rosalind	69,941	0.558	54	
Belinda	75,258	0.622	66	
Puck	86,000	0.762	154	
Miranda	129,400	1.414	472	16.5
Ariel	191,000	2.520	1158	14.4
Umbriel	266,300	4.144	1169	15.3
Titania	435,000	8.706	1578	14.0
Oberon	583,500	13.463	1523	14.2
Neptune				
Naiad	48,000	0.296	54	26
Thalassa	50,000	0.312	80	24
Despina	52,500	0.333	180	23
Galatea	62,000	0.429	150	23
Larissa	73,600	0.554	192	21
Proteus	117,800	1.121	416	20
Triton	354,800	5.877*	2705	13.6
Nereid	5,514,000	360.16	240	18.7
Pluto				
Charon	19,640	6,387	1212	16.8

Appendix 3. Asteroid Data

Asteroid	*Mean distance from Sun, millions of km*	*Orbital period, years*	*Diameter, km*	*Rotation period, hours*	*Magnitude*
1 Ceres	399	4.60	940	9.08	7.4
2 Pallas	366	4.62	525	7.81	8.0
3 Juno	348	4.36	259	7.21	8.7
4 Vesta	339	3.63	576	5.34	6.5
5 Astraea	349	4.13	120	16.81	9.8
6 Hebe	328	3.77	204	7.28	8.3
7 Iris	317	5.51	208	7.14	7.8
8 Flora	304	3.27	162	12.35	8.7
9 Metis	337	3.69	158	5.08	9.1
10 Hygeia	442	5.54	430	17.50	10.2
279 Thule	637	8.23	130	?	15.4
433 Eros	269	1.78	22	5.3	8.3(max)
951 Gaspra	302	3.28	20	20.0	14.1
588 Achilles	718	11.77	116	?	15.3
1566 Icarus	194	1.12	1.4	2.3	
2100 Ra-Shalom	97	0.76	1	19.7	
3200 Phaethon	106	1.27	5	?	
2060 Chiron	1278	50.68	150	?	15
5145 Pholus	3050	93	150	?	17

Appendix 4. Comet Data

Comet	*Period, years*	*Orbital eccentricity*
Encke	3.3	0.85
Grigg-Skjellerup	5.1	0.66
D'Arrest	6.2	0.66
Giacobini-Zinner	6.5	0.71
Finlay	6.9	0.70
Faye	7.4	0.56
Tuttle	13.8	0.82
Crommelin	27.9	0.92
Tempel-Tuttle	32.9	0.90
Halley	76.1	0.97
Swift-Tuttle	130	0.98

Appendix 5. Meteor Showers

Meteor Shower	*Begins*	*Max.*	*Ends*	*ZHR**	*Parent comet*
Quadrantids	1 Jan	4 Jan	6 Jan	60	—
Lyrids	19 Apr	21 Apr	25 Apr	10	Thatcher
Eta Aquarids	24 Apr	5 May	20 May	35	Halley
Delta Aquarids	15 July	6 Aug	20 Aug	20	—
Perseids	23 July	11 Aug	20 Aug	75	Swift-Tuttle
Orionids	16 Oct	22 Oct	27 Oct	25	Halley
Taurids	20 Oct	3 Nov	30 Nov	10	Encke
Leonids	15 Nov	17 Nov	20 Nov	var.	Tempel-Tuttle
Geminids	7 Dec	13 Dec	16 Dec	75	Phaethon (asteroid)

*ZHR = Zenithal Hourly Rate. This is the number of meteors which would be expected to be seen with the naked eye by a single observer under ideal conditions, with the radiant at the zenith. As these conditions are never fulfilled, the observed rate is always appreciably less than the theoretical ZHR.

Appendix 6. The Constellations

* = invisible, or always very low, from New Zealand and much of Australia and South Africa.
** = circumpolar, or almost so, from these areas. The column headed 'Culmination' gives the approximate date at which the constellation is at its highest in mid-evening.

Name	*English name*	*Culmination*	*Chart*	*Page No*	*1st- magnitude star(s)*
Andromeda	Andromeda	23 Nov	23	168	—
Antlia	The Air-Pump	10 Apr	14	147	—
Apus**	The Bird of Paradise	5 July	1	120	—
Aquarius	The Water-bearer	9 Oct	18	156	—
Aquila	The Eagle	30 Aug	21	162	Altair
Ara	The Altar	25 July	16	150	—
Aries	The Ram	14 Dec	23	168	—
Auriga*	The Charioteer	4 Feb	5	132	Capella
Boötes	The Herdsman	16 June	20	160	Arcturus
Caelum	The Graving Tool	15 Jan	3	128	—
Camelopardalis*	The Giraffe	6 Feb	7	134	—
Cancer	The Crab	16 Mar	11	143	—
Canes Venatici*	The Hunting Dogs	22 May	10	141	—
Canis Major	The Great Dog	16 Feb	3	128	Sirius
Canis Minor	The Little Dog	28 Feb	11	143	Procyon
Capricornus	The Sea-Goat	22 Sept	18	156	—
Carina**	The Keel	17 Mar	9	137	Canopus
Cassiopeia*	Cassiopeia	23 Nov	—	—	—
Centaurus**	The Centaur	14 May	16	150	Alpha Centauri, Agena
Cepheus*	Cepheus	13 Nov	—	—	—
Chamaeleon**	The Chameleon	15 Apr	1	120	—
Circinus**	The Compasses	14 June	16	150	—
Columba	The Dove	1 Feb	3	128	—
Coma Berenices*	Berenice's Hair	17 May	15	149	—
Corona Australis	The Southern Crown	14 Aug	17	155	—
Corona Borealis	The Northern Crown	3 July	20	160	—
Corvus	The Crow	12 May	14	147	—
Crater	The Cup	26 Apr	14	147	—
Crux Australis**	The Southern Cross	12 May	1	120	Acrux, Beta Crucis
Cygnus	The Swan	13 Sept	21	162	Deneb
Delphinus	The Dolphin	14 Sept	21	162	—
Dorado**	The Swordfish	31 Jan	8	136	—
Draco*	The Dragon	8 July	21	162	—
Equuleus	The Foal	22 Sept	21	162	—
Eridanus**	The River	25 Dec	8	136	Achernar
Fornax	The Furnace	17 Dec	24	169	—
Gemini	The Twins	19 Feb	6	133	Pollux
Grus	The Crane	12 Oct	19	158	—
Hercules	Hercules	28 July	22	165	—
Horologium	The Clock	25 Dec	8	136	—

Name	*English name*	*Culmination*	*Chart*	*Page No*	*1st- magnitude star(s)*
Hydra	The Watersnake	29 Apr	13	145	—
Hydrus**	The Little Snake	10 Dec	1	120	—
Indus	The Indian	26 Sept	19	158	—
Lacerta*	The Lizard	12 Oct	23	168	—
Leo	The Lion	15 Apr	12	144	Regulus
Leo Minor*	The Little Lion	9 Apr	12	144	—
Lepus	The Hare	28 Jan	3	128	—
Libra	The Scales	23 June	17	155	—
Lupus	The Wolf	23 June	16	150	—
Lynx*	The Lynx	5 Mar	7	134	—
Lyra*	The Lyre	18 Aug	21	162	Vega
Mensa**	The Table	28 Jan	1	120	—
Microscopium	The Microscope	18 Sept	19	158	—
Monoceros	The Unicorn	19 Feb	3	128	—
Musca**	The Fly	14 May	1	120	—
Norma	The Rule	3 July	16	150	—
Octans**	The Octant	(Polar)	1	120	—
Ophiuchus	The Serpent-bearer	26 July	22	165	—
Orion	Orion	27 Jan	2	125	Rigel, Betelgeux
Pavo**	The Peacock	29 Aug	19	158	—
Pegasus	The Flying Horse	16 Oct	23	168	—
Perseus*	The Perseus	22 Dec	7	134	—
Phoenix**	The Phoenix	18 Nov	19	158	—
Pictor**	The Painter	30 Jan	9	137	—
Pisces	The Fishes	11 Nov	23	168	—
Piscis Australis	The Southern Fish	9 Oct	18	156	Fomalhaut
Puppis	The Poop	22 Feb	9	137	—
Pyxis	The Mariner's Compass	21 Mar	9	137	—
Reticulum**	The Net	3 Jan	1	120	—
Sagitta	The Arrow	30 Aug	21	162	—
Sagittarius	The Archer	21 Aug	17	155	—
Scorpius	The Scorpion	18 July	17	155	Antares
Sculptor	The Sculptor	10 Nov	23	168	—
Scutum	The Shield	15 Aug	17	155	—
Serpens	The Serpent	21 July	22	165	—
Sextans	The Sextant	8 Apr	11	143	—
Taurus	The Bull	14 Jan	4	130	Aldebaran
Telescopium	The Telescope	24 Aug	17	155	—
Triangulum Australe	The Southern Triangle	7 July	16	150	—
Tucana**	The Toucan	1 Nov	19	158	—
Ursa Major*	The Great Bear	25 Apr	10	141	—
Ursa Minor*	The Little Bear	(Polar)	—	—	—
Vela	The Sails	30 Mar	9	137	—
Virgo	The Virgin	26 May	15	149	Spica
Volans	The Flying Fish	4 Mar	9	137	—
Vulpecula	The Fox	8 Sept	21	162	—

ASTRONOMICAL SOCIETIES

The national societies are: The Royal Astronomical Society of New Zealand (Astronomy Centre, Carter Observatory, Wellington); the Astronomical Society of Australia; and the South African Astronomical Society. In New Zealand there is a planetarium at the Carter Observatory, and one is planned for the Auckland Observatory. In South Africa there are planetaria in Johannesburg and Cape Town.

Local societies include:

New Zealand
Auckland Astronomical Society
Canterbury Astronomical Society (Christchurch)
Ashburton Branch, CAS
Dunedin Astronomical Society
Foxton Beach Astronomical Society
Gisborne Astronomical Society
Hamilton Astronomical Society
Hawera Astronomical Society
Hawkes Bay Astronomical Society (Napier)
Nelson Branch, Royal Astronomical Society of New Zealand
New Plymouth Astronomical Society
Northern Astronomical Observers Association (Auckland)
North Otago Astronomical Society (Oamaru)
Palmerston North Astronomical Society
Palmerston North Boys High School Astronomical Society
Rotorua Astronomical Society
South Canterbury Astronomical Society (Timaru)
Southland Astronomical Society (Invercargill)
Taranaki Active Astronomers Group
Wairarapa Astronomical Society, (Masterton)
Wanganui Astronomical Society
Wellington Astronomical Society
Whakatane Astronomical Society
Whangarei Astronomical Society

Australia
(New South Wales)
Astronomical Society of New South Wales
Sutherland Astronomical Society
Western Sydney Amateur Astronomical Society
Illawarra Astronomical Society
Hawkesbury Astronomical Association
Northern Districts Society of Amateur Astronomers
Port Macquarie Astronomical Association
Taree Astronomical Society
The Astronomical Society of the Hunter (Tighes Hill, NSW)
Shoalhaven Astronomers
Parkes Astronomy Club
Astronomical Association of the Central Coast (Toukley, NSW)
Astronomical Society of Coonabarabran
Canberra Astronomical Society

(Queensland)
Astronomical Association of Queensland (Brisbane)
Southern Astronomical Society (Pimpama)
Brisbane Astronomical Society
South-East Queensland Astronomical Society (Dayboro, Queensland)
Townsville Astronomy Group
Bundaberg Astronomical Society
Sun Coast Astronomical Society (Koffat Beach, Queensland)
Cairns Astronomy Group (Caravonica, Queensland)

(South Australia)
Astronomical Society of South Australia
Bowman Park Astronomical Society

(Tasmania)
Astronomical Society of Tasmania (Launceston)

(Victoria)
Astronomical Society of Victoria (Melbourne)
Astronomical Society of Frankston
Astronomical Society of Geelong
Albury Wodonga Astronomical Society
Ballarat Astronomical Society
La Trobe Valley Astronomical Society (Churchill)
The Bendigo District Astronomical Society

(Western Australia)
Astronomical Society of Western Australia (South Perth)
Murdoch Astronomical Society
Astronomical Society of the South-West (Bunbury, WA)
Goldfields Astronomical Society
Newman Astronomical Society
Pilbara Astronomical Society

(Northern Territory)
Astronomical Society of Alice Springs

South Africa
The South African Astronomical Society has branches in Cape Town, Johannesburg, Durban and Port Elizabeth.

A BRIEF GLOSSARY

Absolute magnitude. The apparent magnitude which a star would have if it could be viewed from a standard distance of 10 parsecs (32.6 light-years).

Altazimuth mounting. A mounting upon which a telescope can be swung freely both in *altitude* (up or down) and *azimuth* (east or west).

Aphelion. The furthest distance from the Sun of an orbiting planet, comet or other body.

Apparent magnitude. The apparent brightness of a celestial body. The lower the magnitude, the brighter the body.

Asteroids. Minor planets.

Astronomical unit. The distance between the Earth and the Sun: 149,597,893 km (in round figures 150 million km).

Aurora. Polar lights; aurora australis (southern), aurora borealis (northern). They are phenomena of the upper atmosphere, due to charged particles emitted by the Sun.

Azimuth. The bearing of an object in the sky, from north (0 degrees) through cast, south and west.

Binary star. A system made up of two stars which are physically associated, and are orbiting round their common centre of gravity.

Black hole. A region round a very condensed body from which not even light can escape.

Celestial sphere. An imaginary sphere surrounding the Earth, whose centre coincides with that of the Earth's globe.

Cepheid. A short-period variable star whose period of fluctuation is linked with its real luminosity.

Chromosphere. That part of the Sun's atmosphere lying above the bright surface or photosphere.

Circumpolar star. A star which never sets.

Conjunction. (1) A close approach of two bodies in the sky — i.e. a line of sight effect. (2) Inferior conjunction of a planet within the Earth's orbit occurs when the planet is virtually between the Earth and the Sun. (3) Superior conjunction occurs when the planet is on the far side of the Sun relative to the Earth.

Corona. The outermost part of the Sun's atmosphere.

Cosmology. The study of the universe considered as a whole.

Counterglow. English name for the Gegenschein.

Culmination. The maximum height of a celestial body above the horizon.

Declination. The angular distance of a celestial body north or south of the celestial equator.

Dechotomy. Exact half-phase of the Moon or an inferior planet.

Earthshine. The faint luminosity of the night side of the Moon, due to light reflected on to the Moon from the Earth.

Eclipse, lunar. The passage of the Moon through the Earth's shadow.

Eclipse, solar. The temporary blotting-out of the Sun by the interposition of the Moon.

Ecliptic. The apparent yearly path of the Sun among the stars.

Equator, Celestial. The projection of the Earth's equator on to the celestial sphere.

Equinoxes. The two points where the ecliptic cuts the celestial equator.

Extinction. The apparent dimming of a celestial body because of the absorption of its light in the Earth's atmosphere.

Faculae. Bright, temporary patches above the Sun's photosphere.

Flares, solar. Brilliant eruptions in the outer part of the Sun's atmosphere.

Galaxies. Systems made up of stars, nebulae and interstellar matter.

Gamma-Rays. Very short-wavelength radiations.

Gegenschein. A faint glow exactly opposite to the Sun in the sky. Like the Zodiacal Light, it is due to interstellar matter being lit up by the Sun.

Gibbous phase. The phase of the Moon or a planet between half and full.

Hertzsprung-Russell (HR) diagram. A diagram in which the stars are plotted according to their spectra and their luminosities.

Inferior planets. Planets which are closer to the Sun than we are, i.e. Mercury and Venus.

Infra-red radiation. Radiation with a wavelength longer than that of visible light.

Kiloparsec. One thousand parsecs (3260 light-years).

Libration. The apparent 'tilting' of the Moon as seen from the Earth, due to the Moon's changing orbital speed.

Light-year. The distance travelled by light in one year: it is equal to 9.4607 million million kilometres.

Local group. A group of more than two dozen local galaxies. It includes the Milky Way system, the Magellanic Clouds and the Andromeda Spiral.

Lunation. The interval between successive new moons: 29d 12h 44m.

Magnetosphere. The region round a celestial body in which the magnetic field of that body is dominant.

Main sequence. A band on the HR Diagram from top left to bottom right.

Megaparsec. One million parsecs.

Meteor. Cometary debris; a tiny particle which burns away in the Earth's upper air.

Meteorite. A body which comes from the asteroid belt and lands on Earth. Meteorites may be irons, stones or a mixture of both.

Nebula. A cloud of gas and dust in space.

Neutrino. A fundamental particle with little or no mass and no electrical charge.

Neutron star. The remnant of a very massive star which has exploded as a supernova. Neutron stars are small and incredibly dense.

Newtonian reflector. A telescope in which the light is collected by a curved mirror, and sent to the eyepiece via a smaller flat mirror in the upper part of the tube.

Nova. A star which suddenly flares up; the outbreak takes place in the highly-evolved component of a binary system.

Occultation. The covering-up of one celestial body by another.

Opposition. The position of a planet or other body in the sky when it is exactly opposite to the Sun.

Orbit. The path of a celestial body.

Parallax, trigonometrical. The apparent angular shift of an object when viewed from two different directions.

Parsec. The distance from which a body would show an annual parallax of one second of arc: 3.26 light-years, 206,265 astronomical units or 30.857 million million kilometres.

Penumbra. (1) The area of partial shadow to either side of the main cone of shadow cast by the Earth. (2) The lighter part of a sunspot.

Perigee. The position of the Moon or other body in orbit round the Earth, when closest to the Earth.

Perihelion. The position of a body orbiting the Sun when at its closest to the Sun.

Perturbations. The disturbances in the movements of a celestial body caused by the gravitational pulls of other bodies.

Phases. The apparent changes in shape of the Moon and planets due to the differing amount of sunlit side turned Earthward.

Photosphere. The bright surface of the Sun.

Planetary nebula. A small, hot, dense star surrounded by a shell of gas. It is not truly a nebula, and has nothing to do with a planet!

Poles, celestial. The north and south points of the celestial sphere.

Position angle. The apparent direction of one object with reference to another, measured from north through east, south and west.

Prominences. Masses of glowing hydrogen rising from the Sun's surface. They were formerly, and quite incorrectly, termed Red Flames.

Proper motion, stellar. The individual motion of a star on the celestial sphere.

Quadrature. The position of the Moon or a planet when at right-angles to the Sun as seen from Earth.

Quasar. A very remote, super-luminous object, now thought to be the core of a very active galaxy. (Also known as a QSO or Quasi-Stellar Object.)

Radial velocity. The toward-or-away movement of a star relative to the Earth.

Radiant. The point in the sky from which the meteors of any particular shower appear to come.

Retrograde motion. Orbital or rotational motion in the sense opposite to that of the Earth. (I have likened it to the movement of a car going the wrong way round a roundabout!)

Right ascension. The angular distance of a celestial body from the vernal equinox.

Scintillation. Star-twinkling. It is due entirely to the effects of the Earth's atmosphere.

Sidereal period. The revolution period of a body round the Sun or a planet (also known as *orbital period*).

Solar wind. A flow of atomic particles streaming outward from the Sun.

Solstices. The times when the Sun is at its maximum distance (23½ degrees from the celestial equator around 22 June (northern) and 22 December (southern)).

Superior planets. All planets lying further from the Sun than we are.

Supernova. A colossal stellar outburst. (Type I) the total destruction of the white dwarf component of a binary system. (Type II) the result of the explosion of a star of high mass. Type II supernova produce neutron stars, which when detectable at radio wavelengths are known as *pulsars*.

Tektites. Small, glassy objects found in localized areas; not now generally believed to come from the sky.

Terminator. The boundary between the day and night hemispheres of the Moon or a planet.

Transit. (1) The passage of a celestial body across the observer's meridian. (2) The projection of Mercury or Venus against the disk of the Sun.

Twilight. The state of illumination after sunset when the Sun is less than 18 degrees below the horizon.

Vernal equinox. The position where the Sun crosses the celestial equator, moving from south to north.

White dwarf. A small, very dense star which has exhausted its nuclear 'fuel'.

Zenith. The observer's overhead point.

Zenithal hourly rate (ZHR). The number of meteors which would be expected to be seen by an observer under ideal conditions, with the radiant of the shower at the zenith.

Zodiac. A belt stretching round the sky, 8 degrees to either side of the ecliptic, in which the Sun, Moon and planets (except Pluto) always lie.

Zodiacal light. A cone of light rising from the horizon and extending along the ecliptic. It is due to thinly-spread interplanetary material being lit up by the Sun.

Index